The Great Paradox of Science

The Great Paradox of Science

Why Its Conclusions Can Be Relied Upon Even Though They Cannot Be Proven

MANO SINGHAM

OXFORD
UNIVERSITY PRESS

OXFORD
UNIVERSITY PRESS

Oxford University Press is a department of the University of Oxford. It furthers
the University's objective of excellence in research, scholarship, and education
by publishing worldwide. Oxford is a registered trade mark of Oxford University
Press in the UK and certain other countries.

Published in the United States of America by Oxford University Press
198 Madison Avenue, New York, NY 10016, United States of America.

Library of Congress Cataloging-in-Publication Data
Names: Singham, Mano, author.
Title: The great paradox of science : why its conclusions can be relied
upon even though they cannot be proven / Mano Singham.
Description: New York : Oxford University Press, [2020] |
Includes bibliographical references.
Identifiers: LCCN 2019021305 | ISBN 9780190055059 (hardback) |
ISBN 9780190055066 (updf) | ISBN 9780190055073 (epub)
Subjects: LCSH: Science—Methodology. | Science—Philosophy. |
Knowledge, Theory of.
Classification: LCC Q175 .S57225 2020 | DDC 501—dc23
LC record available at https://lccn.loc.gov/2019021305

1 3 5 7 9 8 6 4 2

Printed by Sheridan Books, Inc., United States of America

To my daughters Dashi and Ashali

and

my grandsons Thomas and Benjamin

Contents

The Great Paradox of Science

Introduction

The great appeal of science has been its undoubted success in enriching our lives not only in practical ways but also in showing how over time things that on the surface once seemed so inexplicable became understandable, and how vastly diverse phenomena are unified by being revealed to be based on a few underlying principles. All these seemed so fascinating to me from my early teens that I could not imagine a better way of spending my life than studying science more deeply. The world of physics, with its logical structure and underlying mathematical elegance, seemed to promise unlimited frontiers for a lifetime of fascinating investigation.

But upon graduating from high school in Sri Lanka, my hopes for entering university to pursue a physics degree lay in serious doubt because I had failed to pass a language requirement. Disappointed, I looked for a backup career to make a living and went into accountancy. During the training program, it turned out that I did quite well, mainly because the mathematics and logic involved came easily to me. I was quite comfortable with double-entry bookkeeping and could distinguish assets from liabilities and debits from credits. But my heart was not really in it. So when I overcame the language barrier at the last minute and thus qualified for university, I was elated. I went to the head of the accountancy school and told him that I was dropping out to pursue a physics degree. He tried to dissuade me, saying that he thought I was making a mistake and that I had a gift for accountancy. He then added what he must have thought was the clinching argument. He said that with accountancy, there is a fixed body of knowledge and that with diligent study one could eventually have the satisfaction of having mastered all of it. But when it came to science, one could never achieve that state and would always be left with unanswered questions. He thought I would find that extremely frustrating.

That well-meaning educator did not realize that he had said exactly the wrong thing. I can see why the prospect of a never-ending search for new knowledge, and the idea that one might never reach the goal of knowing everything in one's field of study, might be unsettling for some. But for me, that was the main allure of science, to seek and find answers to interesting and important problems, but yet never run out of fascinating questions to explore.

But after completing my undergraduate studies and then pursuing graduate work toward a doctoral degree in theoretical physics, it seemed like my idea that physics provided unlimited frontiers for exploration might be mistaken. The last quarter of the twentieth century saw one spectacular success in physics after another that raised the possibility that we were getting close to uncovering the fundamental particles that make up the universe and the underlying laws that govern their behavior. There was even talk of finding the "theory of everything" and even of "the end of science."

Similar talk had emerged a century earlier, before the physics revolutions in relativity and quantum mechanics in the first few decades of the twentieth century shattered those expectations and opened up radically new ways of viewing the world. But hubris is part of human nature and tempts us to think that this time things are different, that we have finally got it right, and are not prone to the same errors as our predecessors. So while the scope of scientific knowledge is now so vast that no single individual can know everything, as the head of the accountancy school seemed to think was possible in his field, it seemed like as a *community* of scientists we were approaching a time when we would know all there is to know, at least in their broad outlines, with only mopping-up operations remaining. We seemed tantalizingly close to uncovering the ultimate truths about the nature of the universe.

I had mixed feelings about this. Scientists are puzzle-solvers at heart and while there is something exhilarating about sensing that one is close to cracking open a difficult puzzle and arriving at a solution, achieving such success, like coming to the end of an ingenious mystery novel, also brings with it a sense of anti-climax, a wistful feeling of "Is that all there is?" and the wish for more. The thought that future generations of scientists would not have the same excitement of tackling major open questions brought with it a tinge of sadness, similar to the sentiment expressed by eminent physicist Paul Dirac in 1939 that "In 1926 it was possible for people who were not very good to solve important problems, but now people who *are* very good cannot find important problems to solve" (Livio 2013, 159).

But after obtaining my doctorate and anticipating playing my own small role in what might possibly be the twilight of science, I stumbled upon the book *The Structure of Scientific Revolutions* by Thomas Kuhn (Kuhn 1970) that looked more deeply at the history and philosophy of science. I was startled by Kuhn's claim that while the progress of science was undeniable, there was no reason to think that there was any final frontier at all that science was progressing toward, let alone that it was a finite distance away and that we were close to it.

My prior ignorance of the work of Kuhn and other philosophers of science is not surprising. The formal study of the history and philosophy of science does not form part of the curriculum in science graduate programs. Instead, what scientists acquire is *folklore* about the nature of science that practicing scientists share amongst themselves and pass on to their students. This folklore is then spread to the general public via popular books, articles, and talks by scientists.

My curiosity was piqued by the fact that there was a vast discrepancy between that folklore and what historians, philosophers, and sociologists of science were uncovering. This resulted in my pursuing two parallel tracks of study, physics on the one hand and philosophy of science on the other. In so doing, I became increasingly convinced that philosophers of science were shedding important light on the nature of science that needed to be better known both by scientists and the general public (Okasha 2016).

The major question that came to the fore in my investigations was how it could be that science seemed to be progressing so rapidly and the knowledge it produced so successful in revolutionizing our lives if it was not, as so many of the philosophers claimed, approaching something that we could call "the truth."

While deeper knowledge about the nature of science is not essential for the actual *practice* of scientific research, which is why scientists have done very well without bothering too much about it, I have two other, much more practical, goals for this book as well. The first is that the methods and reasoning by which scientists arrive at their conclusions would enable people, if they became familiar with it, to make much better decisions in all aspects of their lives, what we can call "practical rationality." The second arises from the fact that science has a huge impact on public policy, and when we enter that area of intersection, the misconceptions, lack of understanding, and outright distortions about the logic and nature of science become increasingly significant. I am concerned that the lack of general awareness of how scientists arrive at their knowledge and why that knowledge is so reliable and has proven to be so powerful has enabled those who have agendas that go against the scientific consensus, such as those groups opposed to vaccinations or the teaching of evolution or who are climate change skeptics or who market questionable products, to sow confusion and doubt and prevent meaningful action that can save lives and the planet. These groups use anecdotal evidence or cherry-pick data or rely on people who are not credible experts to advance their causes, whereas *the reliability of science arises because of the creation of consensus conclusions by credible experts using comprehensive bodies of evidence that are systematically acquired and evaluated using scientific logic and must pass through institutional filters such as legitimate peer-reviewed publications.*

Successfully combating powerful groups that seek to undermine the scientific consensus on important issues requires much more than knowledge of the folklore of science because the more sophisticated members of those groups exploit that folklore, with its shaky epistemology, to their advantage. Supporters of science need to understand the weaknesses of their folkloric understanding of science and then go beyond them and reach deeper levels, as this book seeks to do, in order to better combat the false narratives of science's opponents. As philosopher of science H. M. Collins wrote:

So long as scientific authority is legitimated by reference to inadequate philosophies of science, it is easy for laymen to challenge that authority. It is easy to show that the practice of science in any particular instance does not accord with the canons of its legitimating philosophies. The fears of those who object to relativism on the grounds of its anarchic consequences are being realized, not as a result of relativism, but as a consequence of an over-reliance on the very philosophies that are supposed to wall about scientific authority. Those walls are turning out to be made of straw. If new walls are to be constructed, they will have to have their foundations laid in scientific practice—in our understanding of the role of the tacit elements of scientific expertise, and the way this expertise, not a philosophical system, can give justification to an opinion about the natural world. (Collins 1983, 99–100)

If we are to more effectively counter the misunderstandings and distortions, some deliberately fostered, that surround public understanding of science, a deeper understanding of the logic of science is required and this necessitates coming to grips with profound questions of proof, theories, laws, and how we establish the existence and nonexistence of entities.

I have written this book to address both philosophical and practical needs. My book seeks to help build a firmer foundation for science by taking the conclusions of the philosophers, historians, and sociologists of science seriously and addressing the resulting paradox of how scientific theories can be so successful in explaining the world around us despite the lack of any assurance that those theories are true or have an increasing level of correspondence with reality. I argue that those two requirements are unnecessary for understanding the success of science and rather than buttressing its credibility, are actually a hindrance and a distraction because they raise metaphysical questions that cannot be resolved, where I use the word "metaphysical" not in its original sense of first principles or ultimate causes but in the more common and slightly pejorative sense of being abstruse and undecidable using standard methods of reasoning. Furthermore, the new understanding of science that will replace the folkloric knowledge will provide people and policymakers with better tools to make sound rational decisions on matters that they encounter in their everyday lives and in public policy areas, using the same methods that scientists use to arrive at judgments on important questions.

This book is aimed at those people who are interested in science even if they have little or no formal training in it, but practicing scientists will, I hope, also gain a deeper understanding of the underlying knowledge structure of their own work. It starts by dispelling many of the myths and misconceptions and folklore surrounding the nature of scientific knowledge, thus laying the groundwork for why we need a deeper understanding of how science arrives at its knowledge structures and why we are justified in having such trust in them. This leads to the formulation in chapter 20 of what I refer to as the Great Paradox, the fact that despite what many believe, the success of science need have little to do with truth or correspondence with an objective reality. The last two chapters provide a model for the resolution of that paradox.

In order to guide the reader, I will start by laying out the outline of the argument in the book, and each chapter summary will be reproduced at the beginning of each chapter to aid in following the argument.

The structure of the book

Chapter 1 introduces the main themes of the book and argues that it is science that enables us to go well beyond the world that we can access purely via our senses, but that doing so requires the use of advanced equipment and technology to gather data and deep inferential reasoning to extract useful information from that data.

Chapter 2 looks at how popular accounts of scientific history tend to be viewed through the lens of present-day science, focusing largely on those developments that led to the current state and presenting that history as a more-or-less linear process toward that end. By largely ignoring all the cross-currents and confusion that are almost always present in science, the resulting accounts tend to be seriously distorted and should be treated with a great deal of skepticism, even though they can serve useful pedagogical purposes.

Chapter 3 examines popular misconceptions about the nature of science. The notion of scientific truth as correspondence with reality is argued to be not necessary to do science but is helpful in communicating the ideas of science amongst scientists and the general public. Scientists are always seeking what *works* and thus tend to be philosophical and methodological opportunists, quite willing to abandon one approach and shift to an alternative if they think that it will produce better results.

In chapter 4, I discuss how science investigates phenomena that lie outside our direct sensory experience and the difference in the logical arguments used to infer the existence of entities from those used to establish nonexistence. Applying the same logic and reasoning that scientists have used to establish the nonexistence of many things would enable people to rid themselves of many unsupported and superstitious beliefs.

Chapter 5 deals with how despite the counterintuitive nature of many scientific conclusions, it is because science works so well that people accept them. While the problem of induction prevents us from generalizing from a few instances and predicting the future purely by what we have observed in the past, it is the belief that the laws and theories of science are the best that we have and getting better with time that gives us confidence in their predictions. The importance of having a better understanding of the way that the scientific community uses the words "law," "theory," "hypothesis," and "fact" is emphasized.

Part Two of the book begins with chapter 6, in which a detailed case study of the search for the answer to the question of how old the Earth is serves as a paradigm for how scientific "facts" need not be unchangeable. The age has oscillated wildly before settling on the currently accepted value of 4.54 billion years, with religion, politics, and other nonscience factors influencing the search along the way. The final consensus involves a complex interplay of theories from geology, biology, physics, chemistry, and paleontology.

In chapter 7, I discuss how the way we arrived at the current age of the Earth shows that reversals of seemingly firm conclusions are the *norm* in science, not the exceptions, and form an integral part of its process. They are thus not a cause for alarm and the community of scientists has over time developed ways to arrive at consensus judgments that, while not infallible, can command considerable confidence.

Part Three of the book begins with chapter 8, which takes a historical look at the search for true knowledge and how scientific consensus conclusions have changed from being thought of as unchanging and infallible to now being considered as just *provisionally* true, the best we have at any moment. Establishing the validity of scientific propositions has become so difficult that it is now the preserve of a few specialists who have the time, resources, and expertise to carry out the required investigations.

Chapter 9 deals with the importance for *all of us* to understand the epistemology of science in order to better counter those who selectively use such knowledge to advance dangerous anti-science agendas, such as arguing that science is just another species of opinion and requires faith. While doubt is always present in science, even in the absence of absolute certainty we can still have a high degree of confidence in scientific consensus judgments.

Chapter 10 addresses some popular misconceptions about the nature of scientific theories. It discusses why they cannot be proven true or false, that scientific revolutions always involve at least a *three*-cornered struggle involving at least two theories and experiment, and that experimental data are never sufficient to uniquely determine a theory. Although we have not been able to specify both necessary and sufficient criteria to distinguish science from nonscience, there do exist necessary conditions that any scientific theory and law must satisfy, and that is that they must be both *naturalistic* and *testable*.

Chapter 11 expands upon the point touched on in the previous chapter of how scientific theories are so deeply interconnected that they cannot be investigated in isolation, and how this prevents individual theories from unequivocally being proven true or false and creates difficulties when choosing between two competing theories.

Chapter 12 looks at how we have a natural propensity to invent theories all the time based on our experiences, and it is the testing of these theories and their refutation and replacement with new theories that more accurately represents scientific practice.

Part Four of the book begins with chapter 13, which looks at what we can learn from axiomatic systems and the role of proofs in arriving at truths. It discusses why there are limits to what we can prove to be true even in mathematics because we cannot construct a framework that is both *complete* and *consistent* for any nontrivial system. Science is slightly different in that we deal with *quasi-axiomatic* systems with the additional element of experimental data or observations that we can compare with the predictions of theories, but it faces the same problem.

Chapter 14 looks at how scientific logic has strong similarities to the way that the legal system uses logical arguments and evidence to arrive at judgments. The logic used depends on whether a proposition is an existence claim or a universal one, and this determines where the burden of proof lies. That same kind of logic is used also in everyday life, though many people may not consciously realize that they are doing so.

Chapter 15 looks at mathematical proofs that use the method of *logical contradiction* and examines how far this can be taken in science. While the existence of any entity can never be proven by this method, the *nonexistence* of certain entities can.

Chapter 16 looks at the important role that *negative* evidence, the things we do *not* observe, plays in scientific logic. This is illustrated by the example of why we so strongly believe that only two kinds of electric charges exist, to the extent of basing our entire modern technology on it, even though we have not proved it to be so, and indeed cannot even hope to do so.

Chapter 17 uses what we have learned about scientific logic to evaluate the status of four theories that are currently at the frontiers of physics and command much attention in the media: dark matter, dark energy, string theory, and the multiverse.

Part Five of the book begins the process of weaving together all the earlier threads, with this chapter looking at what historians, philosophers, and sociologists of science have uncovered about the way that the scientific community chooses between competing theories and how the process proceeds rationally and systematically even if questions of truth are not determinative.

Chapter 19 looks at why achieving consensus in science can be slow and getting unanimity of views on some scientific questions is almost impossible, because those who are determined to find ways to preserve their beliefs can always find ways to do so.

Chapter 20 finally confronts the Great Paradox: If science is progressing, what could it possibly be progressing *toward* if not the truth? It argues that the way that scientific paradigms evolve is analogous to the process of biological evolution, in that both are conditional on the environment that exists at any time and thus there is no reason to believe that the evolution of scientific theories is heading toward a unique truth. The strong impression of directionality is because scientific history in textbooks is reconstructed *after the fact*. Scientific evolution, like biological evolution, is not teleological.

Chapter 21 builds on Charles Darwin's metaphor of the Tree of Life to construct two other tree metaphors that illustrate the nonteleological nature of science. One is the Tree of Scientific Paradigms that exemplifies the process by which scientific paradigms evolve, with new ones emerging over time that have resulted in their present variety and diversity. The other is the Tree of Science that represents the evolution of scientific knowledge *as a whole*.

Chapter 22 uses the Tree of Science to resolve the paradox of how scientific theories can work so well and be so successful in explaining the world around us despite the lack of any assurance that those theories are true, that they are even approaching something that we can call the truth. It argues that the ideas of truth and correspondence with reality are unnecessary for understanding the success of science and are actually a hindrance and a distraction, because they raise metaphysical questions that cannot be resolved.

PART ONE
WHY UNDERSTANDING THE NATURE OF SCIENCE IS IMPORTANT

1

Did dinosaurs have tea parties?

[This chapter introduces the main themes of the book and argues that it is science that enables us to go well beyond the world that we can access purely via our senses, but that doing so requires the use of advanced equipment and technology to gather data and deep inferential reasoning to extract useful information from that data.]

A friend of mine recounted to me a question that had popped into his mind when he had been visiting the Grand Canyon. The park ranger was telling the group the importance of not littering by specifying how long different kinds of trash last, saying that aluminum cans would take as long as 200 years to decompose, plastic bottles would take 450 years, glass bottles would take the longest at about a million years, and so on. What struck my friend was that these times, although long by our human time scale, were small when compared to geological time scales.

It occurred to him that dinosaurs went extinct about 65 million years ago. We think of many of the dinosaurs as impressive in their size and the way they dominated the world in their time, roaming freely over the Earth with everything as their prey and with few predators to fear. But we don't associate them with any *culture*. We don't associate them with discovering fire or building homes or creating artifacts such as pottery and tools for their use.

But my friend wondered how we *know* that they didn't do any of these things. Could it be that they were actually more advanced than we give them credit for and did at least some of those things but that all the evidence has disappeared over the long time since they were wiped out? It is true that we have not discovered any artifacts dating back to the same period as the dinosaurs that show signs of conscious design. But once all the dinosaurs went extinct, anything they created would start to decompose, with most returning at various times to their elemental forms. Is there anything at all that could and perhaps should have survived until the present day that would suggest that dinosaurs had some sort of culture? We think they didn't because we have not found anything other than their fossils. But can we be certain about this? The process of fossilization occurs when, under suitable conditions, water infiltrates the pores and cavities in bone, wood, and shell, and minerals in the water replace the organic matter and become hardened. These stony fossils are the things that last almost indefinitely, but dinosaur artifacts may not have been amenable to that process.

Let's pose the question in another way. Human beings have created a vast number of artifacts that cover the surface of the globe. If all of us, like the dinosaurs, were to some day disappear, relatively suddenly due to some catastrophe or slowly because we failed to properly protect our environment, would at least some traces of our civilization last forever or would there come a time when there would be no sign that we lived lives that went beyond mere existence? If extraterrestrials were to arrive on Earth hundreds of millions or even billions of years after our disappearance, would they find only human fossils like we now find with dinosaurs or would they also uncover evidence that a sophisticated civilization once existed?

It seems hard to imagine that there would be no traces at all of our civilization, given the extent of the things that we have created. But it may well be the case, since there does not seem to be anything that lasts for more than a few million years or so before decomposing into elemental forms or becoming buried deep underground due to plate tectonics. If so, then why could it not be the case for dinosaur artifacts too? After all, humans have been around for a mere two million years (and modern humans only for 200,000 years) and thus produced all these things in a much shorter period than the dinosaurs who roamed the Earth for around 150 million years. Why do we believe that dinosaurs did not do anything at all during that time other than eat, sleep, and reproduce? Why could it not be that they too developed some kind of society, however rudimentary, whose traces have disappeared in the 65 million years that have elapsed since they went extinct? We can immediately think of some obvious hindrances to such advancements, such as their lack of opposable thumbs, but can we be sure that such deficiencies completely rule out any kind of culture?

The idea of dinosaurs having a culture seemed so preposterous that my friend was too embarrassed to pose this question to the park ranger in the Grand Canyon. But actually it is a good question because trying to answer it requires coming to grips with the entire structure of *deep inferential knowledge*, of how we know things that are not directly accessible via our senses. Our human senses have extremely limited ranges. Gaining knowledge of deep time (i.e., going back to the biological origins of life on Earth or the physical origins of the Earth and the universe) or of deep space (the outer limit of the visible universe) or the world of the very small (such as the electron) or the very fast (such as light) lies outside the abilities of any individual. It can only be obtained indirectly and needs the collective expertise of many people, often using extremely long chains of subtle reasoning and advanced mathematics applied to data gleaned from complicated instruments. Science uses precisely such methods to arrive at much of its knowledge, and has been extremely successful in utilizing the answers obtained. Science is what we turn to in trying to answer important questions about phenomena that lie outside the range of direct human experience and deciding how sure we are of that knowledge.

There are those who use the deep inferential nature of scientific knowledge to argue that it is less sound than other forms of knowledge that are more readily and transparently accessible. But in fact, apart from those things that we personally experience, inferential knowledge is pretty much all that we have, except that some forms of it seem, on the surface, to be more concrete and reliable than others. For example, take records of historical or even contemporary events. We may think that because they were recorded by eyewitnesses or by people taking the testimony of eyewitnesses, those records can be considered as factual. But we know that eyewitness testimony and memories can be notoriously unreliable and that recorders, even if they are diligent and sincere in their attempts at being accurate, must necessarily be selective about what they document, and can find it hard to avoid subjective factors coloring their accounts. This is why we have long-lasting debates about historical events even when there are contemporaneous accounts by supposed eyewitnesses. What we think of as incontrovertible, objective, historical facts are often the *consensus verdicts* of a few people who have analyzed the raw data and produced a narrative that is plausible and supported by evidence. As consumers of such accounts, we are usually unable to do an independent analysis and need to rely upon the credibility of those experts for our "facts."

Almost all knowledge consists of this kind of deep inferential knowledge. The knowledge that we obtain from science is different only in degree from the knowledge obtained in other fields, in that the inferential chains are often longer, more abstract in nature, and require more esoteric tools to uncover. Scientific knowledge also seems much more unfamiliar because it is usually exploring a world that we cannot imagine easily because it lies beyond the reach of our senses, or because it deals with events that took place a long time ago before humans were even around to record them, or in distant parts of the universe that we can never hope to get to. Those who seek to pursue agendas that go against the scientific consensus in some areas have tried to exploit this unfamiliarity of the deep inferential knowledge structure of science, to sow doubt as to the validity of conclusions about phenomena that are so far removed from our immediate experience.

One group of people who have adopted the most extreme forms of this skepticism are the religious fundamentalists who believe, based on their interpretation of religious texts, that the Earth has existed for less than ten thousand years and that the theory of evolution is false. One such sect instructs the children of its followers to challenge teachers who make assertions about anything that predates recorded history by asking them "Were you there?," implying that only those things that have been directly witnessed by humans are the things we can know for sure and that everything else is of doubtful validity (Patterson 2015). Astrologers, psychics, faith healers, and others of that ilk assert the existence of supernatural and other processes that they claim are beyond the ability of science to investigate.

But religious fundamentalists and believers in the supernatural are not the only ones who have sought to find ways to challenge the robustness of scientific conclusions. There are also business interests that have challenged the conclusions of science about the health risks of smoking and the damage caused by acid rain and chlorofluorocarbons. Opponents of childhood vaccinations similarly deny the scientific consensus on the safety, effectiveness, and necessity of vaccines. Climate change denialists challenge the conclusion that the planet is in danger of undergoing irreversible and potentially catastrophic changes by questioning the validity of climate models and the estimates of long ago temperatures. All these groups suggest that some piece of contradictory evidence, or making an ad hoc change somewhere, can decisively refute the consensus conclusions of the scientific community.

These concerted efforts to undermine confidence in scientific conclusions need to be countered because of the harmful effects of the policies that these critics seek to advance. But in doing so we must be cautious and not make claims in support of science that cannot be sustained, because those can be exploited to sow further doubt as to the credibility of science. Some of the critics of science use quite sophisticated arguments from the fields of history and philosophy of science, and countering them requires equal or even greater levels of sophistication about the knowledge generated by those fields.

The success of science is such that surveys find overwhelming majorities in America who say that science has had a mostly positive effect on society and that science has made life easier. Scientists rank third in public esteem, behind the military and teachers but ahead of medical doctors, engineers, clergy, journalists, artists, lawyers, and business executives (Pew Research Center 2009). This support exists in spite of the fact that most people do not really understand how science works. Why should this lack of understanding be a source of concern? After all, many people don't understand how planes fly and yet they board them for long distance travel without any qualms. Many people do not understand how smartphones work and yet use them for all manner of communications. They don't understand how microwave ovens work and yet can use them to cook and heat their food with facility. In living our lives today, we are surrounded by technology that we have little understanding of and yet we are unfazed and use it effectively and easily. What is wrong with treating science the same way, with little or no understanding of its inner workings but using the information and products created by science for our benefit? Why bother to go to all the trouble of trying to understand how all that knowledge was obtained and why it is so reliable?

There are good reasons why achieving more widespread and deeper understanding of the nature of science is beneficial. Very little actually hinges on whether or not we understand how our everyday appliances work. But a lot hinges on understanding the deep inferential reasoning used in acquiring scientific knowledge. Science is not just a collection of factual information that underlies modern technology. Arriving at that information involves making difficult decisions about which scientific theories we can rely on and which should be rejected. Since the consequences of those decisions are so important, over time the community of scientists has developed ways that enable them to make reasoned and reliable judgments. Although this book is mostly concerned with the evolution of science and not technology, the importance of technological advances in spurring scientific curiosity and enabling the investigation of scientific theories cannot be overemphasized.

Many of the ways of making scientific judgments have practical relevance in all aspects of our daily lives and can be enormously helpful in decision-making for individuals and for public policy. Indeed we often unconsciously use many of those scientific decision-making strategies but because we are not explicitly aware that we are doing so, our everyday decision-making is often idiosyncratic and inconsistent. By becoming *explicitly* aware of how the scientific community arrives at its conclusions about which theories can be relied upon and which ones should be jettisoned, we not only gain more confidence in those theories, we become able to make better judgments about important questions concerning our society and even the more mundane aspects of our own lives. But getting to that happy state of a deeper understanding of the way science works is not straightforward.

To really understand how science works, we need to understand the *logic* of science, how the scientific community reasons its way to conclusions. Learning about the principles of logic in science is important because one needs a common framework in order to adjudicate disagreements. A big step toward resolving arguments lies in either agreeing to a common framework by which a judgment can be arrived at or deciding that one cannot agree and that further discussion is pointless in the absence of new information. Either outcome is more desirable than going around in circles endlessly, not recognizing what the ultimate source of disagreement is.

For example, one important recurring issue in science is how we decide on the existence or nonexistence of postulated entities. Why do we believe that electrons exist but that the aether does not? Establishing existence seems the more straightforward of the two possibilities. For example, we believe in the existence of horses because there is direct evidence for them. But establishing nonexistence is more problematic. We (or at least many of us) do not believe in the existence of unicorns, leprechauns, pixies, dragons, centaurs, mermaids, fairies, demons, vampires, werewolves, and a host of other mythical entities because there is no credible evidence for them *even though we cannot incontrovertibly prove they do not exist*. What makes us so sure of their nonexistence?

This book will explain how we know what we know about what exists and what does not, and argue that even though we often cannot *logically* prove nonexistence, that does not prevent us from treating many things as *effectively nonexistent* on the grounds that postulating their existence is unnecessary. While this basic idea is simple, complications arise when the evidence is not in a form that is directly accessible via our senses and thus available to everyone, but instead requires using sophisticated equipment to obtain the data and complex theories involving deep inferential reasoning in order to interpret and understand the evidence. This results in most people not being able to evaluate the evidence themselves, and then the question of what constitutes credible evidence becomes more problematic. But if people understand the process by which the evidence is evaluated by credible experts who have the knowledge and skills to do so, this might help in resolving many of the doubts and uncertainties that have been exploited by some to advance dubious agendas.

How we make judgments about the existence or nonexistence of any entity is based on scientific logic, but this logic is not only applicable in the province of science. It is used in all manner of academic disciplines that deal with the empirical world. Indeed, it is also what everyone uses in everyday life while not being explicitly aware that they are doing so. This book will try to make explicit what has long been implicit. But scientific logic applies to far more than questions of existence and nonexistence of entities. It is also used to determine the laws that govern the behavior of entities and how we judge whether these laws are valid, and is thus worth knowing about for those reasons alone.

The ability of science to answer difficult questions seems self-evident. People have got so used to science working so well in providing the underpinnings of the marvels of our modern society that it seems that it can only do so because it is generating true knowledge that tells us what the world is really like. But how valid is that belief? This book will argue that the idea that we can know if scientific theories are true or that they provide an increasingly accurate representation of reality cannot be sustained upon closer examination. When one looks closely at the structure of scientific knowledge, the initial easy assurance that it works so well because its theories must be true becomes harder to sustain, and this is what makes its success so surprising. But as will become clearer, such a claim turns out to be unnecessary and even undesirable because it cannot be proven. The basis for why science works so well has to be sought elsewhere.

The claim that we have no reason to think that the theories of science are true or that they are approaching the truth in terms of increased correspondence with reality will stop some readers cold because it seems to make the immense success of science and its progress over the last few centuries inexplicable. This is the central paradox that this book seeks to address. It can be expressed more concretely by posing the following question: If extraterrestrial beings were to visit the Earth at some point, that would imply they possessed technology superior to ours. If we could communicate with them, would we find that the theories of science on which their technological prowess was based were the same or a close approximation to ours (as would be the case if the theories of science are true or approaching truth), or would we find them to be completely different, suggesting that there is no unique truth out there waiting to be discovered?

People who believe the former will sympathize with the view expressed by philosopher of science Peter Kosso when he writes, "The explanatory and predictive success of theories [of science] would be an absolute miracle if the theories were not true," and that the truth of theories can be gauged by the degree of coherence that exists between all the theories that go into explaining phenomena (Kosso 1992, 100, 128). But incredulity is a shaky guide when we are dealing with deep inferential knowledge. After all, it was not that long ago that the marvelous complexity of the living things in the world were taken as obvious proof of an intelligent designer at work, a view that some hold even today. But we now know that we can understand that complexity without invoking any supernatural designer.

Others who are more familiar with the literature on the history, philosophy, and sociology of science may be disturbed for a different reason. It is quite common for those who seek to undermine the credibility of science to indulge in the deplorable practice of "quote mining," to take out of context statements that scientists make and misleadingly use them. Supporters of science may fear that these critics will seize on the assertion that scientific theories are neither true nor approaching truth to argue that the conclusions of science cannot be trusted and that this implies the truth of their own pet alternative theories. The critics will ignore this book's arguments that we *never* have such a binary choice, where either one or a competing theory is true. While scientific theories may not be true in any absolute sense, it is an indisputable fact that they do work exceedingly well. The alternative theories proposed by critics not only have no plausible claim to being true, they perform abysmally when compared with science. Hence it should be undeniable that scientific theories are preferable to the alternatives. In defending their value we need not, and should not, resort to truth claims that cannot be substantiated, because those ultimately weaken the case for science. But reaching that level of deeper understanding of why science works so well takes a little effort. To understand how science can be so successful without being true, the Great Paradox that the title of this book refers to requires us to go deeper into the nature of science, beyond the folklore and simple truth claims.

My hope is that this book will serve as a guide to navigating that terrain. It necessarily starts with dismantling the popular idea that truth is a determining factor in how scientific theories evolve and then proceeds in the last part to present the resolution of the resulting paradox. The goal for this book is to make clearer to the general public, whatever their level of familiarity with and expertise in science, how scientists make judgments about what exists and does not exist and what theories to keep and what to reject. In doing so, I will take as a given what most scientists and nonscientists alike believe about the world, that what our senses and devices detect gives us information about an objectively existing "real" world and that we all share a common reality. I postulate such a realistic metaphysics because that is what most people who are not professional philosophers of science believe to be the case, and that is the audience to whom this book is addressed. The problem is that even with that assumption, we cannot know when we have grasped that reality or be assured that we are getting closer to it, so reality cannot be determinative of how science progresses. (This book will not be going into deeper philosophical questions, such as whether such an objective reality indeed exists and whether our senses and measuring devices are reliable detectors of that reality. That discussion can be found elsewhere in the philosophical literature. For an elementary introduction on quantum mechanics and objective reality, see Singham 2000; Ball 2018.)

This book has been strongly influenced both directly and indirectly by Charles Darwin's groundbreaking work on the theory of evolution *On the Origin of Species*. Just as he said that his entire book consisted of "one long argument" about the nature of evolution (Darwin 1859, 459), so is the present book one long argument about the nature of science. Along the way, while exploring all these issues, I will be dealing with many minor puzzles and paradoxes of science. That journey will take us into areas of knowledge involving science, mathematics, law, history, philosophy, sociology, and epistemology of science. Whether readers agree with my argument or not, I hope they will learn some new and interesting things along the way.

Much of the material will be familiar to philosophers, historians, and sociologists of science, but not so much to practicing scientists or the general public. This may be partly due to the fact that much of the literature written by members of those three disciplines and the language they use, even in popular presentations, may be unfamiliar to outsiders. This book seeks to bridge that gulf and my hope is that my background as a practicing scientist will make my presentation of that material more accessible. In doing so, I have tried to follow the dictum attributed apocryphally to Albert Einstein that "Everything should be made as simple as possible, but not simpler" (Calaprice 2005, 290) by shifting material that may be more difficult to a Supplementary Materials section at the end.

As to the specific questions of a possible ancient dinosaur culture and whether a future extraterrestrial visitation long after our own extinction would discover any signs of our own presence, I will not be going into them, fascinating though those questions are, especially in the light of recent tantalizing suggestions that the brains of dinosaurs may have been larger and more developed in complexity than formerly surmised (Brasier et al. 2014). Once the reader has finished this book, I will leave it to her to speculate as to what might be looked for in order to be able to provide more definitive answers to those questions.

2

The traps of scientific history

[This chapter looks at how popular accounts of scientific history tend to be viewed through the lens of present-day science, focusing largely on those developments that led to the current state and presenting that history as a more-or-less linear process toward that end. By largely ignoring all the cross-currents and confusion that are almost always present in science, the resulting accounts tend to be seriously distorted and should be treated with a great deal of skepticism, even though they can serve useful pedagogical purposes.]

Any attempt to examine the logic of science necessarily involves delving into the historical development of science in order to provide examples of how the scientific community arrives at judgments about which theories they think are useful and thus should form the basis of future research, and which theories should be discarded. As we go through life and the education process, we acquire an informal understanding of that decision-making process. However, if one happens to take a detour and looks at scholarly works on the history, sociology, and philosophy of science, one finds that those understandings will not survive. Any close examination of the actual sequence of events that have accompanied major scientific developments will quickly convince the reader that the descriptions of the nature of science found in textbooks, popular scientific magazines and books, and the news media should be treated as largely folklore and viewed with deep skepticism.

This folklore often follows the "heroic" model, in which a famous scientist such as Nicolaus Copernicus or Isaac Newton or Charles Darwin or Albert Einstein, aided by a crucial experiment or observation or insight, makes a major discovery that changes the way we view the world. Such narratives usually take the form of specifying who discovered what and when and how, and suggest that the piecemeal addition of such discoveries eventually led to the creation of the powerful mosaic of scientific knowledge that we currently have. One is given the impression that our knowledge of the world is inexorably increasing, getting better with time, providing us with an increasingly accurate and comprehensive understanding of how the world works. It seems to be only a matter of time before all major questions will be answered and only mopping-up operations remain, heralding the "end of science" (Horgan 1996).

In reality, it turns out that even what seems like the simple task of pinning names and dates to major scientific discoveries is usually a futile exercise.

Statements of the form "This scientist discovered this phenomenon (or proposed a new theory) on this date" turn out to be at best misleading and at worst wrong. Scientific discovery is not something that is easily pigeonholed in such a manner. When one examines any historical event up close one finds a cacophony, with multiple points of view, confusion, vacillations, reversals, and inconsistencies that over time slowly disappear as one viewpoint becomes dominant.

This poses a problem for the historian as to how best to approach the topic. A regimen of dispassionate fact-gathering and scrupulous descriptions of events without imposing any theoretical perspective on them would, even if that were possible, result in amassing a wealth of detail that on its own may shed little light on the process of scientific discovery, though it may provide valuable material for others to use as raw material for analysis. More common is the approach in which one uses the present state of knowledge as a guide to the interpretation of the past, a method sometimes referred to as "presentism." But this poses a danger in that if one is not careful, one can end up distorting history by emphasizing just those elements that led to the present state and largely ignoring those that did not.

In the 1930s, historian Herbert Butterfield described this tendency in the context of political history in his book *The Whig Interpretation of History*, which examined how Martin Luther and the Protestant Reformation are often portrayed by historians.

> The tendency in many historians [is] to write on the side of Protestants and Whigs, to praise revolutions provided they have been successful, to emphasize certain principles of progress in the past and to produce a story which is the ratification if not glorification of the present. (Butterfield 1931, v)
>
> . . .
>
> The whig interpretation of history is not merely the property of whigs and it is much more subtle than mental bias; it lies in a trick of organisation, an unexamined habit of mind that any historian may fall into. It might be called the historian's 'pathetic fallacy.'
>
> It is the result of the practice of abstracting things from their historical context and judging them apart from their context—estimating them and organising the historical story by a system of direct reference to the present. (Butterfield 1931, 30–31, quotation in original)

When carried to an extreme, such manipulative historical treatments are pejoratively referred to as "whiggish histories" that, as pointed out by Larry Laudan, can lead to unfair characterizations of those who happened to hold views that are no longer seen as viable.

To ignore the time-specific parameters of rational choice is to put the historian or philosopher in the outrageous position of indicting as irrational some of the major achievements in the history of ideas. Aristotle was not being irrational when he claimed, in the fourth century B.C., that the science of physics should be subordinate to, and legitimated by, metaphysics—even if that same doctrine, at other times and places, might well be characterized as irrational. Thomas Aquinas or Robert Grosseteste were not merely stupid or prejudiced when they espoused the belief that science must be compatible with religious beliefs. (Laudan 1977, 131)

While these passages by Butterfield referred to political history, he later expanded the domain and added that histories of science often fall prey to that same temptation, in which the contributions of past scientific activity and individual scientists are evaluated mainly by how much they contribute to the state of current knowledge (Butterfield 1957). In this later work, Butterfield tried to provide a fuller picture by looking at the major scientific revolutions associated with the heliocentric solar system, the laws of inertia, dynamics and gravitation, the circulation of blood, phlogiston and the chemical revolution, and evolution. He described the many cross-currents of views, personalities, and social and political contexts that surrounded each period of transition in order to explain the major intellectual hurdles that had to be overcome before the final theory became dominant. He says that minimizing this complexity, ignoring all the avenues *not* taken, and focusing on just the few individuals whose ideas now strike us as modern, is to seriously distort our understanding of scientific history, because we end up painting a picture of inevitable and unrelenting progress (Butterfield 1957, viii–ix).

Sociologist of science Barry Barnes expands on what whiggish history looks like.

Whig history treated the institutions of earlier generations as incompletely constructed versions of our own: their beliefs as partial representations of what is now fully understood; their innovations, whether in custom, social organization, technique or natural knowledge, as movements towards the more 'advanced' forms found today. Historical change was preconceived as 'progress', and accounted for as a movement closer to the present. It was almost as though the present was a cause of historical change, pulling the past in conformity to it by some magnetic attraction, or perhaps pre-existing like some genetic code in the developing social organism, telling it the final perfected form into which it had to grow. (Barnes 1982, 4, quotations in original)

The reader may immediately recognize many examples of this approach in popularizations of science written by scientists and journalists. It is easier to succumb to the temptation of writing whiggish history in the sciences because the current state of scientific knowledge is seen as the *inevitable* outcome of the process, because it is either the true one or, even if not true in any absolute sense, is better than what we had before, is the best of all possible alternatives that are available to us now, and lies on the path that will take us to the truth. There is a widespread and firm belief that science is *progressing* by getting *better* all the time despite occasional detours due to error. When science is viewed in this way, to evaluate the past purely in terms of how much it contributes to the present can be seen as more justifiable, even desirable, than in the case of political history or indeed the history of almost any other field.

The eminent biologist Ernst Mayr argued precisely along these lines, saying that the history of science differs from political history in significant ways and that Butterfield erred in seeing both as similar, because science clearly demonstrates progress in ways that politics does not. Mayr's view is that while in political revolutions the final outcome could have gone in many directions depending on contingent factors such as the power and persuasiveness of the competing factions, science *builds on past successes* and the winner is the theory that is unquestionably *better* than the one that it replaced and thus we are always improving. Mayr argues that because of this difference, a developmental approach is unavoidable in writing science history, unlike in political history, where it is merely a choice.

> My own conclusion is that Butterfield was ill advised in his literal transfer of the whig label from political history to history of science. It was based on the erroneous assumption that a sequence of theory changes in science is of the same nature as a sequence of political changes. Actually the two kinds of changes are in many respects very different from each other. In political changes succeeding governments often have diametrically opposed objectives and ideologies, while in a succession of theories dealing with the same scientific problem each step benefits from the new insights acquired by the preceding step and builds on it. Galileo, indeed, had a superior understanding of physics than the Greeks, Newton than Galileo, and Einstein and modern physicists than Newton. The same is true for the sequence Linnaeus-Lamarck-Darwin-modern evolutionary biology or, for that matter, for any historical sequence of scientific theories. For this reason the historiography of science proceeds by necessity in many respects very differently from political historiography. This is most clearly recognized by those who write developmental history of science. (Mayr 1990)

Mayr thinks that historians of science are pursuing an impossible ideal when they try to avoid the charge of whiggishness by simply reporting factual events and amassing details about phenomena without seeking or imposing any narrative arc that is derived from the present. While he acknowledges that whiggish histories can be found in science, he argues that this occurs when the historian gives short shrift to, completely ignores, or treats contemptuously those theories of the past that were in contention at one time but were rejected and no longer form part of the scientific consensus. He argues that if those losing points of view are given, within the constraints of time and space, a respectful hearing that shows clearly why they are wrong, then a developmental history is valid.

Mayr's point of view is common in scientific circles and reflects the strong sense of confidence in the scientific community of their ability, at any given point in history, to make correct judgments about which theories are right and which are wrong, or at least which theories are better and which are worse, so that the history of science can be viewed as on a positive path, always improving our knowledge, except for brief detours due to error that are later rectified. Since most scientists tend not to be professional historians, they are vulnerable to the temptation to indulge in whiggish history not only because it provides a simpler narrative that is easily digested but also because it gives new generations of students an optimistic view of science and the human condition, in which we are ever marching onward and upward.

Some scientists recognize that the history they are describing is not the whole story even if they don't know what that full story is. For example, Richard Feynman in his book *QED* outlines the origins of one of the most successful theories of physics known as quantum electrodynamics, but then cautions:

> What I have just outlined is what I call a "physicist's history of physics," which is never correct. What I am telling you is a sort of conventionalized myth-story that the physicists tell to their students, and those students tell to their students, and is not necessarily related to the actual historical development which I do not really know! (Feynman 1985, 6)

But not all scientists are as aware as Feynman that they are merely propagating myths and not the full story. As we well know, if you repeat a "myth-story" enough times and it gets passed on from generation to generation, it can acquire the status of a fact, and popularizations of science by scientists are often the way that these "facts" gain widespread acceptance.

It can be argued that in science, propagating whiggish history can in a narrow sense be an actual advantage. In the following extended passage from his classic work *The Structure of Scientific Revolutions*, Thomas Kuhn describes why this might be so, using the idea of a "paradigm" to denote a framework based on a theory within which scientific work is carried out.

The temptation to write history backward is both omnipresent and perennial. But scientists are more affected by the temptation to rewrite history, partly because the results of scientific research show no obvious dependence upon the historical context of the inquiry, and partly because, except during crisis and revolution, the scientist's contemporary position seems so secure, More historical detail, whether of science's present or of its past, or more responsibility to the historical details that are presented, could only give artificial status to human idiosyncrasy, error, and confusion. Why dignify what science's best and most persistent efforts have made it possible to discard? The depreciation of historical fact is deeply, and probably functionally, ingrained in the ideology of the scientific profession, the same profession that places the highest of all values upon factual details of other sorts. Whitehead caught the unhistorical spirit of the scientific community when he wrote, "A science that hesitates to forget its founders is lost." Yet he was not quite right, for the sciences, like other professional enterprises, do need their heroes and do preserve their names. Fortunately, instead of forgetting these heroes, scientists have been able to forget or revise their works.

The result is a persistent tendency to make the history of science look linear or cumulative, a tendency that even affects scientists looking back at their own research. For example, all three of Dalton's incompatible accounts of the development of his chemical atomism make it appear that he was interested from an early date in just those chemical problems of combining proportions that he was later famous for having solved. Actually those problems seem only to have occurred to him with their solutions, and then not until his own creative work was very nearly complete.

...

Or again Newton wrote that Galileo had discovered that the constant force of gravity produces a motion proportional to the square of the time. In fact, Galileo's kinematic theorem does take that form when embedded in the matrix of Newton's own dynamical concepts. But Galileo said nothing of the sort. His discussion of falling bodies rarely alludes to forces, much less to a uniform gravitational force that causes bodies to fall. By crediting to Galileo the answer to a question that Galileo's paradigm did not permit to be asked, Newton's account hides the effect of a small but revolutionary reformulation in the questions that scientists asked about motion as well as in the answers they felt able to accept. (Kuhn 1970, 138–40)

In her exhaustive analysis of how during the first half of the twentieth century what we call the "Copenhagen interpretation" of quantum mechanics became dominant, Mara Beller shows that the people who are now seen as instrumental in its formulation (Niels Bohr, Werner Heisenberg, Max Born, and Wolfgang Pauli) were constantly changing their views, at various times adopting ones that contradicted each other and their own earlier views, sometimes opportunistically doing so in order to appeal to different audiences and win converts. Decades later, their reminiscences of past events were revised to create a linear narrative that conformed to their final views. Despite the fact that there were, and still are, viable alternatives to the Copenhagen interpretation, those other voices have largely disappeared from the discussion (Beller 1999).

What is more accurate than a statement of the form "This scientist discovered this phenomenon (or proposed a new theory) on this date" is to say something along the lines of "Before date X, the theory or phenomenon was not part of the scientific consensus but after date Y, it was considered well established." In the intervening period, ideas related to it bounced around among various researchers working on the problem, continually getting refined in the process until they settled into a rough consensus. The assignment of credit and dates for the final formulation of these efforts is an act of *retrospective consensus judgment* by the community of scientists in the field, mostly based on perceptions of who contributed the most to the final formulation and assigning their names to them, along with dates, as easily-identifiable markers. This verdict it is not always unanimous and sometimes bitterly contested by those who feel their contributions have not been adequately recognized.

A detailed anthropological study of daily scientific practice in the Salk Institute laboratory in La Jolla, California shows how difficult it is to assign credit and dates. Bruno Latour describes how scientists arrived at the elucidation of TRF (thyrotropin-releasing factor), a hormone that is secreted by the brain. He says that his observations led him to "the conviction that a body of practices widely regarded by outsiders as well organized, logical, and coherent, in fact consists of a disordered array of observations with which scientists struggle to produce order" (Latour and Woolgar 1979, 36). The final consensus formulation of TRF (i.e., when it became a scientific "fact") emerged some time between January 1968 and January 1970 (Latour and Woolgar 1979, 181–82). Half of the Nobel Prize in Medicine for 1977 for this discovery was shared between the Salk group and another competing group in New Orleans (the other half went to Rosalind Yalow for the techniques she developed that was used in these studies), but each of these two competing groups strongly felt that they deserved sole credit and that the other did not deserve even a share of the award.

But despite these known shortcomings, we still continue (as I will also be guilty of sometimes in this book) to use that misleading shorthand description of scientific discovery in the form of "Scientist X discovered Y on date Z" as long as it serves a pedagogical purpose by providing a concise, convenient, and memorable narrative structure of landmark developments in science. But in doing so, we should always bear in mind that what we are passing on is a myth-story and not an accurate history, let alone a complete one.

3

Misconceptions about the methodology and epistemology of science

[This chapter examines popular misconceptions about the nature of science. The notion of scientific truth as correspondence with reality is argued to be not necessary to do science but is helpful in communicating the ideas of science amongst scientists and the general public. Scientists are always seeking what *works* and thus tend to be philosophical and methodological opportunists, quite willing to abandon one approach and shift to an alternative if they think that it will produce better results.]

Just as the simplified history of science that we informally acquire while learning science or about science is misleading when not actually wrong, similar misconceptions exist about the nature of what is commonly referred to "the scientific method." Popular literature (and school science fair projects) suggests a fairly methodical and standardized process consisting of a list of steps that supposedly represent the stages that scientists go through in their work. The sequence consists of posing a research question, conducting background research on the topic, constructing a hypothesis, conducting experiments to test the hypothesis, analyzing the data, drawing conclusions about the validity of the hypothesis, and communicating the results in the form of a scientific paper.

In reality, the process is never so neat and orderly. The various steps listed here indeed all occur and do so repeatedly, but not in any rigid order. Each element is used as needed by scientists in getting an idea refined and tested, and the process can sometimes seem haphazard to the casual observer as the scientist shifts from one step to another seemingly randomly. But that impression is misleading. What the experienced scientist is doing is adopting a heuristic approach that enables her to quickly make decisions about what avenues seem the most promising to explore at any given stage in the investigation. Quite often (as we saw with the Dalton example quoted by Thomas Kuhn in the previous chapter), the precise formulation of the research question, supposedly the starting point of the investigation, is only arrived at the *end* of the process, simultaneously with the completion of the analysis, because it is only then that the scientist sees clearly what is happening.

But in presenting the final results to the community, a convention has evolved in which scientists present a more orderly, straightforward process (Medawar

1964). The words of T. S. Eliot in his poem "Little Gidding" seem particularly apt when applied to how scientists describe their work in their papers.

> What we call the beginning is often the end
> And to make an end is to make a beginning.
> The end is where we start from.

This rewriting of history is not done out of an intent to deceive but because it seems a waste of time to describe the tortuous path involving many wrong turns that preceded arrival at the final result. Scientists are usually willing, if not eager, to share the real story when asked and such discussions often occur in private communications amongst those working on similar problems. Practicing scientists know that the final public presentation in the form of papers and talks almost never reflects the actual process, but the general public may not be so aware.

Similar misunderstandings exist concerning the philosophy of science as well as its *epistemology*, the theory of knowledge that deals with the methods, validity, and scope of science and that tries to distinguish between justified beliefs and opinions. That science works and has worked exceedingly well for many centuries, even millennia, has naturally aroused curiosity as to the reasons for this success and those who work in the areas of philosophy, history, and the sociology of science have undertaken detailed analyses of the scientific enterprise, using individual scientists and scientific events as their objects of study to try to discern any systematic patterns that could shed light on the epistemology of science. While that scholarly work provides valuable insights into the nature of science, it has not achieved widespread dissemination or understanding among scientists and the public.

Many of these scholars have rejected the idea that the methods of science enable us to measure the truth of our theories by the degree of correspondence that those theories have with the "real world." This pessimistic view has been largely driven by the unquestionable fact that over much of history, scientific theories once thought to be true have later been deemed false, or at least radically misleading. As Larry Laudan states:

> Attempts to show that the methods of science guarantee it is true, probable, progressive, or highly confirmed knowledge—attempts which have an almost continuous ancestry from Aristotle until our own time—have generally failed, raising a distinct presumption that scientific theories are neither true, nor probable, nor progressive, nor highly confirmed. (Laudan 1977, 2)
>
> . . .
>
> Determinations of truth and falsity are *irrelevant* to the acceptability or the pursuitability of theories and research traditions. (Laudan 1977, 120)

. . .

What I am suggesting is that we apparently do not have any way of knowing for sure (or even with some confidence) that science is true, or probable, or that it is getting closer to the truth. Such claims are *utopian*, in the literal sense that we can never know whether they are being achieved. To set them up as goals for scientific inquiry may be noble and edifying to those who delight in the frustration of aspiring to that which they can never (know themselves to) attain; but they are not very helpful if our object is to explain how scientific theories are (or should be) evaluated. (Laudan 1977, 126–27)

Some of the scientists who worked in the early days of quantum mechanics in the first half of the twentieth century understood these limitations. This may be because that particular theory forced them to confront the question of whether measurements merely *revealed* the existence of *pre-existing* properties of the world (what is referred to an objective reality and seems almost obviously true) or whether the measured properties *did not exist until they were created by the act of measurement itself* (a highly counterintuitive idea). Thus the question of the existence of an objective reality was starkly posed and was front and center in their attempts to make sense of that theory. The very nature of quantum mechanics forced these scientists to have more modest views on what was the best that science could hope to achieve. Albert Einstein, for example, considered the goal of seeking to reveal reality to be unattainable and that even the concept of reality was meaningless. As he wrote in a letter in 1918:

"The physical world is real." . . . The above statement seems to me intrinsically senseless though, like when someone says: "The physical world is a cock-a-doodle-doo." It appears to me that "real" is an empty, meaningless category (drawer) whose immense importance lies only in that I place certain things inside it and not certain others. . . (Real and unreal seems to me to be like right and left.) (Stachel and Schulmann 1987, 651)

As Mara Beller explains:

For Einstein, the notion of scientific truth is Kantian and holistic: the truth of a scientific statement does not reside in its correspondence with reality but derives from the adequacy of the unified conceptual model to which it belongs (the empirical adequacy of such a model is one of the conditions of its truth; the logical simplicity of its foundation is another). (Beller 1999, 155, citations omitted)

Another quantum mechanics pioneer, Erwin Schrödinger, felt the same way about truth and reality as Einstein did but conceded that in order to communicate with one another and the general public, we needed to use the language of reality.

> Yet an analysis of Schrödinger's writings reveals instead a very sophisticated position, along neo-Kantian lines: the concept of reality "as such," as it objectively exists independent of all human observers, is indefensible, if not downright meaningless. Similarly, Schrödinger fully understood that the correspondence theory of truth can hardly be sustained. Still, the concept of reality, held Schrödinger, is as indispensable in science as it is in everyday life. There is no distinction in principle between a layman's and a scientist's conception of reality—both are regulative constructs, indispensable for mental (and physical) activity. (Beller 1999, 282)

The idea that in order to effectively communicate with one another we sometimes need to use concepts that we do not believe in is not limited to the idea of reality. Within the last fifty years, and especially in the last two decades, the classical idea of free will, that I could have willfully chosen to take an action different from that which I took just a few moments ago, has been seriously undermined (Soon et al. 2008; Blackmore 2005). But giving up the idea of free will is not easy, for both emotional and practical reasons. Describing behavior without using language that has been developed under the assumption of having free will becomes enormously convoluted. For everyday purposes, it is easier to retain language based on free will. As Isaac Beshevis Singer quipped, "We must believe in free will—we have no choice." One could replace "free will" in the quote with "reality" and it would be as accurate, though now lacking the wit of paradox. Concepts of free will and of truth as correspondence with reality are deeply embedded in our language and thinking and enable us to describe actions and events and communicate with one another without convoluted circumlocutions. They can be considered useful fictions.

The important foundational questions that are raised by quantum mechanics about whether an objective reality exists remain controversial and unresolved to this day but have now become somewhat isolated as a narrow specialized field, away from the gaze of the general public and most of the scientific community, and largely the focus of those more interested in questions concerning the meaning of science. This book will not wade into that controversy but will take as a given the common belief that there exists an underlying objective reality, and instead use that to explore another belief that also seems self-evident, that science is on an inexorable march toward uncovering that reality. Challenging that last notion, as many philosophers of science do, is distasteful to many scientists.

As a result, scientists sometimes express contempt for the attempts by these scholars of science to understand the way they work, as can be seen by the sentiment expressed by an unknown wag that "The philosophy of science is just about as useful to scientists as ornithology is to birds" that is frequently quoted with approbation by scientists. A few scientists even express scorn for the field of philosophy in general, seeing it as disconnected from the empirical world and thus of no relevance to their work.

Some well-known scientists of the past who were contemporaneous with Einstein and played key roles in the development of quantum mechanics were not hesitant to state their view that philosophical considerations were not important to their work in any *practical* sense.

> Nor did Niels Bohr consider philosophical analysis heuristically valuable. He did not take seriously considerations of simplicity, elegance, or even consistency (the "epistemic virtues"), holding that "such qualities can only be properly judged after the event." A harsh and crisp verdict came from Paul Dirac: "I feel that philosophy will never lead to important discoveries. It is just a way of talking about discoveries which have already been made." (Beller 1999, 58, citations omitted)

Scientists are not in general driven by explicit philosophical considerations when doing research but all of us, whether we are conscious of it or not, use at least some kind of philosophical framework with which to view and interpret the world. It is impossible to do science without having at least some *implicit* understanding of the philosophy and epistemology of science and most scientists acquire an intuitive idea of it, not through any formal study but incidentally, as a byproduct of the way they learn and do science. Where they tend to use philosophy is when they are seeking to explain and justify their ideas to broader audiences, and in doing so they seize upon whatever philosophical framework meets their immediate needs.

If pressed to explicitly state their conception of the methodology of science, there is one answer that is given frequently by those scientists and nonscientists alike who have put at least some thought into this question, and that is that scientists use the method of *falsification* to make judgments about theories. This approach says that while we cannot prove any theory to be true, we can judge it to be false if its predictions disagree with experiment. This view buttresses the idea that science is progressing steadily toward the truth because it is systematically eliminating false theories. It is the view acquired informally during the scientific training process and has become embedded so strongly in the folklore, and seems to make so much intuitive sense, that many scientists see it as self-evidently true and not as merely one among many possible methodological points of view that one can choose to adopt or discard. As we will see later in chapter 10 (b), when it comes to explaining how science evolves and progresses, the historical evidence just does not support falsification as an adequate explanation.

In sociological studies of the scientific enterprise, scholars observe many schools of thought operating within science, and what is used may vary from scientist to scientist and even from time to time for any individual scientist, depending on the specific needs of the situation. It is only toward the end of their careers, looking back, that scientists may crystallize their lifelong experiences into what seems to them to be a coherent philosophical framework and then impose that view on their reminiscences of the past, as we saw with Dalton in the previous chapter.

For example, in her study of how the Copenhagen interpretation of quantum mechanics came to be dominant, Beller argues that an examination of the contemporary records of the papers and correspondence of the principals involved go counter to the popular belief that the positivist philosophy (that only those things that can be observed or measured had any meaning and should be allowed to enter the theory) always drove the thinking of its advocates like Bohr and Heisenberg.

> Positivist philosophy was less a heuristic principle and more a tool with which theoretical advances could be justified ex post facto. Contrary to the received opinion that Heisenberg's philosophical stand remained stable from the reinterpretation (1925) to the uncertainty paper (1927b), a careful analysis reveals a radical change, if not an about face. . . No coherent philosophical choice between positivism and realism guided Heisenberg's efforts. (Beller 1999, 52)

Albert Einstein, who is widely seen as personifying the quintessential scientist, was more sophisticated in his understanding of scientific epistemology and recognized its value. But even he recognized that scientists are not, and should not be, driven by a single epistemological perspective but should be willing to go wherever experiment dictates and *adopt whatever works* even if that requires shifts in philosophy. In this passage, Einstein describes how science and philosophy intertwine but also draws a distinction between how scientists and epistemologists operate.

> The reciprocal relationship of epistemology and science is of noteworthy kind. They are dependent upon each other. Epistemology without contact with science becomes an empty scheme. Science without epistemology is —insofar as it is thinkable at all —primitive and muddled. However, no sooner has the epistemologist, who is seeking a clear system, fought his way through to such a system, than he is inclined to interpret the thought-content of science in the

sense of his system and to reject whatever does not fit into his system. The scientist, however, cannot afford to carry his striving for epistemological systematic that far. He accepts gratefully the epistemological conceptual analysis; but the external conditions, which are set for him by the facts of experience, do not permit him to let himself be too much restricted in the construction of his conceptual world by the adherence to an epistemological system. He therefore must appear to the systematic epistemologist as a type of unscrupulous opportunist: he appears as a *realist* insofar as he seeks to describe a world independent of the acts of perception; an *idealist* insofar as he looks upon the concepts and theories as the free inventions of the human spirit (not logically derivable from what is empirically given); a *positivist* insofar as he considers his concepts and theories justified *only* to the extent to which they furnish a logical representation of relations among sensory experiences. He may even appear as *Platonist* or *Pythagorean* insofar as he considers the viewpoint of logical simplicity as an indispensable and effective tool of his research. (Einstein 1949, emphasis in original)[1]

In his book *Against Method*, philosopher of science Paul Feyerabend argued that the idea that scientists follow a systematic approach does not explain actual scientific practice and that any examination of actual scientific history will reveal that *scientists broke all the methodological rules all the time* and that any attempt to impose some kind of system would actually paralyze science.

> We find, then, that there is not a single rule, however plausible, and however firmly grounded in epistemology, that is not violated at some time or other. It becomes evident that such violations are not accidental events, they are not results of insufficient knowledge or of inattention which might have been avoided. On the contrary, *we see that they are necessary for progress*. (Feyerabend 1993, 14, my emphasis)

He said that scientists practiced their profession in a spirit of epistemological anarchy that could best be described by the slogan "anything goes." Indeed, one can say that if there is anything that unites scientists, it is the drive to find out by any means at hand what *works*, to arrive at theories that not only explain phenomena but also predict the outcomes of new experiments or observations. And if those predictions are surprising or unexpected and the results confirm them, then scientists are even more willing to accept the theory as being at least provisionally

---

[1] For an entertaining and enlightening exposition on the various philosophical viewpoints toward scientific knowledge that Einstein lists, the reader is encouraged to read Laudan 1990.

true even if it cannot immediately be justified on theoretical grounds and may even be in conflict with other accepted knowledge.

There are some famous illustrations of this. When in 1900 Max Planck proposed his hypothesis that the oscillators in a blackbody cavity had quantized energy levels, there was really no theoretical justification for it other than that it "solved" a long standing problem in blackbody radiation, in that it resulted in a formula that agreed with experimental observations of the blackbody radiation spectrum. Planck himself did not quite believe that his quantum idea represented an underlying reality but thought of it as more of an ad hoc mathematical trick that worked. He thought it served as a placeholder, to be supplanted later by a more well formulated theory, though the great success that it had in explaining blackbody radiation made him later suspect that there may be at least some truth to the idea (Jammer 1966, 22). The same is true for the model proposed by Niels Bohr of the atom as resembling the solar system, with electrons behaving like planets orbiting a central nucleus that serves as the Sun, and the electromagnetic force playing the role of gravity. As with the Planck hypothesis, there was little justification for this model other than that it worked, meaning that it explained some long-standing puzzling features of atomic spectra.

Even though both models had serious drawbacks in that they directly contradicted the expectations of electromagnetic theory, the best theory of science that existed at that time and which is still believed to be true, what prevented the scientific community from rejecting these two seemingly bizarre models out of hand was that despite their significant theoretical drawbacks, they solved major problems of the time that had long resisted the concerted efforts of the leading practitioners in the field. These models also made predictions that (at least partially) turned out to be correct. Scientists pursued and developed those ideas in the hope that they were the precursors to a deeper theory, and they were eventually rewarded by the creation of the more fully formed theory of quantum mechanics that resolved the early contradictions with electromagnetic theory and eventually led to the synthesis that we now call quantum electrodynamics.

The willingness of scientists to abandon strict rigor and conformity to current ideas (at least in the short term) if a new idea seems promising in the long term can also be seen in the way that they use mathematics. Physicists use sophisticated mathematics in the development and articulation of their theories but the way they use it can on occasion cause mathematicians to look askance. This is not because the mathematics used by scientists is wrong but because many mathematical operations are strictly valid only under precisely specified conditions. Mathematicians tend to value rigor highly in their formal proofs and are scrupulous about checking to see if all the conditions hold but scientists are quite willing to overlook such niceties and adopt "quick and dirty" short cuts if they can be made plausible by hand-waving arguments and the end results seem reasonable and agree with experiments. Scientists assume that other researchers will come along later to tidy things up and justify their manipulations. Even Steven Weinberg, one of the more sophisticated users of mathematics in physics, concedes this.

> Since the early nineteenth century, researchers in pure mathematics have regarded rigor as essential; definitions and assumptions must be precise, and deductions must follow with absolute certainty. Physicists are more opportunistic, demanding only enough precision and certainty to give them a good chance of avoiding serious mistakes. In the preface of my own treatise on the quantum theory of fields, I admit that "there are parts of this book that will bring tears to the eyes of the mathematically inclined reader." (Weinberg 2015, 20–21)

Important results were arrived at using such seemingly casual approaches before the calculus of infinitesimals was invented (Bloor 1991, 125–29). Such examples (and there are many more) indicate the commitment of scientists to the attitude that *finding out what works is the most important thing*. What drives all scientists is the desire to find out and keep what does work and discard what does not. In pursuit of that end they will use whatever tools and whatever approaches they find helpful and serves their needs at that moment even if it causes problems elsewhere. They have no choice. In their early days, most new theories will have, like the Planck and Bohr models, contradictions with some features of established knowledge. It usually takes a lot of hard work by large numbers of scientists over a long time to overcome those problems and create a coherent theory that becomes part of the consensus knowledge structure, and even then they are usually not able to completely eliminate all the difficulties. But how the community of scientists decides what works and is promising enough to be worth pursuing more deeply despite initial weaknesses, and what new ideas lack merit and need to be abandoned, lies at the heart of understanding the nature of the scientific enterprise.

Because of this divergence between how science really works and how it is popularly understood to work, initial forays into the study of the nature of science can be unsettling because they can upend the sense of certainty that often pervades our ideas about science, and that resulting uncertainty can extend beyond the processes of science to also its conclusions. It is undoubtedly the case that science forms the basis of the most certain *inferential* knowledge that we have about those things that lie outside the ranges of our direct sensory perceptions and experiences. Science works, and works exceedingly well, in uncovering the secrets of that hidden world, and the argument that this must be because its theories are telling us something true about that world is a seductive one. Some philosophers of science argue that such incredible success would be nothing less than miraculous if our scientific theories were not saying something true, or at least approximately true, about the objective world (Putnam 1975, 73). And yet it turns out that the theories of science that, because of their success, we firmly believe to be true may not be so. There is *always* an element of doubt about all our scientific theories, however robust they may seem on the surface. Despite that, we have every right to be confident enough about many theories that we can treat them *as if we are certain that they are true* and use them as the springboard to acquiring new knowledge, as well as to develop practical applications to which we even entrust our lives.

How can we be so confident about things of which we are not certain, and indeed often cannot be certain, that we can proceed as if we are certain? That is what this book will address. We will see that going beyond superficial understandings and essentially false narratives about the nature of science provides rich rewards to those who seek such deeper understandings and give us greater, not lesser, confidence in its conclusions. But before we get there we have to travel through terrain that may be unfamiliar to some.

Recall the epic adventures of classical literature where the heroes leave the comfort and security of home and head out into an uncertain world where there are enemies to be fought, perils to be faced, dangers and pitfalls that threaten, and fears that must be overcome. When the heroes eventually reach their destination, they are all the more confident for knowing that they have faced their own fears and doubts and uncertainties and overcome them and discovered new things along the way. The intellectual journey into a deeper understanding of science and how we know what we know constitutes a similar experience and can result in a similar range of emotions. We start out with the comfortable sense that science can be trusted because its claims are true and then venture into territory where the basis for that certainty is undermined and everything seems to be in doubt. But for those who stick with the journey and persevere, the reward is that we end up at a place where we have a much deeper understanding of the scientific process and acquire a much more subtle awareness of the meaning of the words "true" and "false" in the scientific context and greater confidence in how scientific judgments are made about what is true and what is false.

Why is this desirable? We need to develop this deeper awareness because it makes us better able to combat the claims of those who would seek to misuse the epistemology of science to undermine conclusions that they find distasteful for various reasons. There are those who are fearful of science because it challenges their beliefs (as in the case of religious objections to the theory of evolution), those who seek to discredit it because its conclusions challenge their economic interests (as in the case of those who deny the scientific consensus on the dangers of smoking, environmental pollution, global warming and climate change), those who have been misled by bad science (as in the case of those who oppose vaccinations against diseases), and those who benefit from people believing in all manner of superstitions (such as astrology). All these groups seek to undermine the authority of science, and its more sophisticated members have probed its epistemology for weaknesses. As a result, some of them are actually more aware of the problems that are discussed in the central section of this book than supporters of science because they are actively looking for its supposed weaknesses, and they use the uncertainties discussed there to advance their cause by exposing the flaws in the folklore of science. To successfully confront them requires at least as deep or deeper understanding of the nature of science.

Those who seek, for whatever motives, to undermine the value of the knowledge produced by science are aided by the fact that the very ubiquity of science's successes can serve to make those successes invisible so that people take them for granted. Unless reminded, people may not fully appreciate that science has revolutionized our lives in ways that no other sphere of knowledge has and thus downplay the significance of its achievements. They are like the character Reg in the Monty Python film *Life of Brian* that is set in the time of the Roman Empire. In his rhetorical effort to inspire his followers to rise up against the Roman occupiers, Reg, attempting to downplay the benefits they provided, asks "What, apart from the aqueduct, sanitation, roads, irrigation, medicine, education, wine, public baths, fresh water system, law and order, public safety, and peace, have the Romans ever done for us?" They are also like Senator Lackavision in the cartoon by Kevin Cannon (see Figure 3.1).

Am I guilty of overdramatization by pitting the pro-science and anti-science camps as being engaged in some kind of epic struggle? I don't think so. Misunderstanding, deliberately misrepresenting, or being ignorant of the nature of scientific inquiry and the way that we arrive at scientific conclusions is not harmless. It can lead to people not being able to distinguish between those conclusions that are reliable and can be depended upon, and a vast array of pseudo-scientific claims that use the language and trappings of science to promote ideas that adversely affect society as a whole and can lead to behaviors and practices that can be harmful to individuals. This has become even more important these days because the Internet has resulted in people being bombarded with pseudo-scientific information from all sides, with little guidance about how to evaluate it. This imposes an obligation on scientists to become more active in combating misinformation and misunderstandings about science (Hoffman 2018). We cannot take for granted that because its merits are self-evident, science can ignore the attacks of those who seek to undermine it. We can and must defend it, and we can better do so if we have a more sophisticated understanding of the goals and methods of science.

Figure 3.1 Credit: Kevin Cannon

I am well aware that some will fear that in arguing against the position that science is creating true knowledge or is at least advancing toward truth, I may be giving aid and comfort to those who would seek to discredit science. But as Steven Shapin says in his book *The Scientific Revolution*, what is being critiqued here is not science but the *stories* we tell about science.

Any account, such as this one, and much recent history and sociology of science, that seeks to portray science as the contingent, diverse, and at times deeply problematic product of interested, morally concerned, historically situated people is likely to be read as *criticism* of science. It may be thought that anyone making such claims must be motivated by a desire to expose science—to say that science is not objective, not true, not reliable—or that such accounts will have the effect of eroding respect for science.

This, in my view, would be both unfortunate and an inaccurate conclusion, Something *is* being criticized here: it is not *science* but some pervasive *stories* we tend to be told about science. Most critics of science tend to be scientists, and I think they are far better placed to do that critical job than historians, sociologists, or philosophers. Science remains whatever it is—certainly the most reliable body of natural knowledge we have got—whether the stories we are told about its historical development and social relations are accurate or inaccurate. Science remains also the most respected component of our modern culture. I doubt very much whether science needs to be defended through perpetuating fables and myths cobbled together to pour value over it. (Shapin 1996, 165, emphasis in original)

So if the idea that the goal of science is to discover those theories that are true or correspond most closely to the real world is not tenable, then what exactly is its goal? That is what the next chapter will discuss.

4
What is the goal of science?

[This chapter discusses how science investigates phenomena that lie outside our direct sensory experience and the difference in the logical arguments used to infer the existence of entities from those used to establish nonexistence. Applying the same logic and reasoning that scientists have used to establish the nonexistence of many things would enable people to rid themselves of many unsupported and superstitious beliefs.]

Scientific research is hard. Reaching the stage of becoming an independent and productive researcher able to pursue one's interests requires a long period of study followed by more years of apprenticeship. Even after that it requires long hours and often tedious detailed work with little chance of obtaining wealth and glory, except for a tiny number who obtain Nobel prizes and the like, and even for them the fame is often fleeting. The respect and approval of the small community of their peers is often their main reward. So why do they do it? Almost all do it because of an intense curiosity, the desire to understand how the world works. Many because it promises to be able to solve important problems and make the world a better place. Yet others go into it because it is intellectually stimulating and rewarding, and has the promise of prestige, a good career, and a modestly comfortable lifestyle. When it comes to individual scientists, there will likely be a range of overlapping motivations for their choice.

But if one had to summarize a key working goal of the *community* of scientists in a very few words, one might suggest that it is to find out what the world is made of, how those constituents interact with one another to produce the complexity that we see all around us, and how to better use that knowledge. What makes up the constituents to be studied can vary from discipline to discipline. For anthropologists and psychologists, it is people that make up the basic units of study. For molecular biologists it may be complex molecules such as DNA and RNA and proteins. For chemists it may be simpler molecules and atoms. For physicists it may be atoms and nuclear particles like protons, neutrons, and electrons, and subnuclear particles like quarks and gluons.

The constraints under which these constituents behave and interact to produce the phenomena being studied are referred to as the *laws* of science (or sometimes *principles*). The explanatory structures that scientists use to describe how the constituents interact are called *theories* and these too will differ according to the discipline. What they all share in common, across all disciplines, is that these theories and laws are believed to be *universal*, in the sense that they work the same way in all places at all times. Given the same set of conditions, they should produce the same outcomes, if not in every case then at least in the statistical aggregate. We call such behavior "law-like" to emphasize that they do not allow for idiosyncratic behavior. To the extent that there seem to be limits of applicability for some of the laws, these limitations too are law-like in the sense that those limitations are unvarying and, provided they are in place, the outcomes should still be predictable. *It is this predictive quality of scientific theories and laws that gives science its power because it enables us to anticipate, control, and influence events. Theories and laws that make no concrete predictions that can be tested are of no use to science.*

But behind all those motivations lies a yet more powerful one and that is the pursuit of truth. A strong belief that drives scientific practice is that there is a real, objective world out there and the ultimate goal is to arrive at a set of complementary and overlapping theories that describe different domains of knowledge but together provide a true description of that entire world. As discussed in the previous chapter, many philosophers and epistemologists have argued that such a goal is unreachable but it is safe to say that most practicing scientists ignore those arguments and carry on regardless, which can be a good thing because the search for true theories is what makes scientists work extraordinarily hard on very difficult problems for long periods of time and has resulted in many spectacular advances. A powerful motivator for scientists is the belief in the existence of a solution for whatever problem they are working on that will reveal truths about the world, and that given sufficient ingenuity, effort, and resources they will find that truth.

There exists a range of viewpoints among scientists as to how far that pursuit of truth can be taken. At one end of the spectrum are the extreme reductionists who believe that a *single* true theory can play the ultimate explanatory role. The reductionist philosophy says that the constituents, laws, theories, and phenomena in one domain are reducible and explicable in terms of those in the domain in which the constituents are smaller. So the behavior of societies can be understood by the characteristics of the people who live in them, the behavior of people can be understood in terms of their biology, biology in terms of chemistry, and chemistry in terms of physics. In this view, everything ultimately reduces to the domain of the smallest constituents of subnuclear particles and so it should not be surprising that the physicists who work in that area are the most attracted to this view of the world since it gives them pride of place. Their goal is to arrive at what some of them call a "Theory of Everything" consisting of a set of equations that would be able to describe all phenomena that have or ever will be observed in the subnuclear world. Since extreme reductionists believe that every other domain is reducible to this domain, at least in principle, the equations governing that world are seen by them as the ultimate theory of the universe.

However, that goal has been dismissed as unrealistic, unattainable, and conceptually misguided by scientists at other points on the size spectrum, who feel that different domains require their own organizing principles. The theories and laws that describe behavior at one size scale of constituents are rarely, if ever, explicable in terms of the constituents, theories, and laws at size scales that are even just one level smaller, let alone at the smallest level, though they might provide some rough insights. While reductionism has undoubtedly proven to be useful within limited ranges of applicability and as a driver in a few areas of scientific practice, it has failed both practically and theoretically within even the limited domain of physics and thus the idea that the explanation for the entire universe can be reduced to a single set of equations governing the behavior of subnuclear particles has been strongly criticized (Laughlin and Pines 2000).

Scientific laws and theories are better described as *emergent* in that they emerge from, and can usually be understood only within, a particular context and except in very rare cases cannot even in principle be reduced to or derived from theories in other domains involving smaller size scales. This is analogous to the way that the behavior of ant and bee colonies display collective behavior that seem purposeful and sometimes give the impression of the entire colony acting as if it were a single intelligent organism, and yet that intelligence or sense of direction is not apparent in the behavior of an individual ant or bee. As another example, the "wetness" property of water is not discernible in any wetness quality of individual water molecules. Similarly, our sense of consciousness and the thoughts generated by our brains are the product of complex neural networks composed of vast numbers of neurons and the even more massive number of connections between them, but those thoughts cannot be split up into components stored in individual neurons. While the *components* of systems are reducible to smaller ones, that is usually not true of the explanatory structures that describe their behavior.

It should be emphasized that the rejection of reductionism does not mean that new concepts emerge magically at the higher size scales. It must be conceded that the concepts at the lower level are ontologically prior to those at the larger scales, in that the emergent properties at the larger constituent size scales are dependent on the properties and interactions at the smaller ones, and can even benefit from some of the insights gained from them. But in practical terms, those insights rarely provide more than purely heuristic benefits because of the rapidly escalating complexities that arise along with the number and size of the constituent entities involved.

Take as a very simple example the motion of two point masses interacting via Newton's force of gravity. This problem can be solved exactly. But add just one more point mass to the mix and you end up with what is known as the "three-body problem," and it was shown back in the nineteenth century that no solution to it is possible in the form of a closed analytic expression. This was partly the reason that while the motion of a single planet around the Sun could be calculated exactly, explaining the stability of the solar system with all the other planets involved is an immensely complicated problem to which we still do not have a definitive answer. (See chapter 10 (b) for more on this.)

In general, one has to develop new concepts, theories, and techniques to generate solutions to the problems for different size scales. But whatever scale we are working at, when it comes to determining the constituents of nature that are outside the range of everyday experience and inaccessible via our senses, we confront the problem of how to establish the existence of these entities and it is here that there is often a misunderstanding amongst the general public about how scientific logic and reasoning works. This is not because the way that scientists establish the *existence* of something is different from the way we establish the existence of things in everyday life, but because the public tends to misunderstand *both* processes. There is an even greater divergence in how we establish the *nonexistence* of things, and the latter is important because when we say in science that some things exist, there is often the unspoken but important corollary that certain other things do *not* exist.

The methods of science allow us to amass evidence in support of the existence of entities that lie outside the range of our senses and draw firm conclusions about their properties. As a result, nowadays most people have no difficulty believing in the existence of atoms and viruses and bacteria and have some understanding of how they behave. But once we have identified the constituents that make up a domain of study, and also feel that the laws that govern their behavior and the theories that explain them are *sufficient* to understand phenomena, we are also *implicitly* saying that there are no other constituents in that domain. After all, what would be the point in saying that something exists but has no impact on the behavior of anything? But the way we arrive at *firm* conclusions about the nonexistence of entities, though important in building up our ideas about the world we live in, tends to not be widely understood because the inferential reasoning that leads to scientific conclusions about the nonexistence of entities is different from that used to infer existence. But nonscientists seem to expect arguments and evidence for establishing nonexistence to be similar to those required for establishing existence, and this may be because most people live in the world where entities lie within ranges that can be detected directly via our senses.

Take for example the question of the existence of vampires or, if you prefer, ghosts or werewolves or zombies or anything else that is familiar because it is part of popular culture and yet has not been conclusively shown to exist. If we adopt popular modes of argument when it comes to discussing something like this, people usually start with two mutually exclusive propositions: It exists or it does not exist. Which one is true? How does one decide? It seems reasonable to assume that both propositions start out on an equal footing and that supporters of each side then marshal evidence and arguments, and whichever one makes the best case carries the day.

What are the arguments and evidence that can be produced in favor of the existence of vampires? Clearly the best one would be to produce a live specimen that has all or most of the properties ascribed to that class of entities and cannot plausibly be classified into any other known category. Although no one has ever produced an actual vampire for examination, over time there has emerged quite a complex theory of their properties based on a vast literature and folklore that encompass myths and legends. These theories involve their need for human blood, aversion to natural light and garlic and crucifixes, their lack of reflected images in mirrors, their ability to make others into vampires by biting them and drinking their blood, and so forth. The level of detail and complexity that has grown over time may be taken by some as prima facie evidence that vampires exist, since most people are hesitant to dismiss widespread and long-standing folklore as being totally false.

Skeptics pointing out that no one has been able to actually produce a vampire specimen prompts defenders to suggest that vampires are ingenious at disguising their presence or may live in locations that no one has looked at or are able to move around stealthily and thus skillfully evade detection. The person arguing against the existence of vampires has a difficult task because the only way to conclusively prove that something does *not* exist is to conduct an exhaustive search of every single location *simultaneously* and come up empty, something that is impossible to do. The best the skeptic can do is debunk specific sightings that can be investigated to show that the claims are false or have alternative, nonvampire explanations. For example, some of the symptoms associated with vampire-like behavior may be due to other causes, such as people having rabies or the disease porphyria (Lane 2002). But this particular debunking method leaves untouched those claims concerning events that occurred long ago or left no tangible traces of evidence.

We could conduct a similar thought exercise for ghosts. There are people who are convinced that certain locations are haunted by the spirits of the dead. As evidence, they point to reports of sightings of ethereal beings and strange unexplained noises in certain locations. And as with vampires, theories have emerged about the properties of ghosts, such as their immaterial nature, occasional partial transparency, and their ability to pass through solid objects and make sounds. Although some of these claims contradict well-established scientific principles (how can an immaterial object make sounds and be partially visible since the production of sound and light waves require material sources?), it is impossible for those skeptical of such claims to prove them false unless one can catch people in the act of creating hoaxes. And even though hoaxes have been exposed many times, believers can still claim that other sightings that have not been debunked are genuine.

The best argument that the naysayer of vampires and ghosts can make is to demand that in order to be believed, a vampire or ghost must be produced for examination and that in the absence of their existence being thus incontrovertibly established, nonexistence is the *default* stance that one must adopt. Strictly applying that rule, vampires and ghosts must be confined to the realm of fiction because if they had been definitely shown to exist, that would have revolutionary implications for science. Such vampires would have been exhaustively studied for their peculiar biology and proof of ghosts would mean that everyone would believe in the existence of an afterlife.

But the believer can argue that even though vampires and ghosts have not been definitely proven to exist, given the ubiquity and longevity of these stories and our inability to *prove* nonexistence, there must be at least some truth to their existence even if all the details about them are not true, and that the inability to produce a real live vampire or ghost for examination is due to our inability to command their behavior. Aphorisms such as "There's no smoke without fire," "It is better to be safe than sorry," or the Shakespearean "There are more things in heaven and earth, Horatio, than are dreamt of in your philosophy," are often used to justify such speculative beliefs. This kind of argument is invoked even more strongly when it comes to the existence of gods, a topic that will be discussed in chapters 15 and 18.

The lack of conclusive ways to convince people of the nonexistence of vampires and ghosts may be put down to the fact that these kinds of entities are not confined to any specific geographical location and, not being related to known things, can be postulated to have properties that can be modified at will to explain away the lack of evidence for their existence. Thus in the popular mind, the result of this debate may be seen as being inconclusive and, given the extraordinary popularity of vampires (and ghosts and zombies) in literature, TV, and films, some people might well believe that they actually exist. Such beliefs can be dangerous. As recently as October 2017, there appeared news reports of vigilante killings of five people in Malawi who were suspected of being vampires (Haltiwanger 2017).

But even well-defined entities that are confined to a reasonably restricted region and not supernatural, such as the Yeti or Bigfoot, can command support among believers despite the lack of convincing evidence in their favor. Consider the case of the Loch Ness monster that supposedly lives in a large lake in Scotland and that believers think is the descendant of a long line of plesiosaurs, a kind of marine reptile that originated about 200 million years ago. While the first reported sightings of this creature date back to medieval times in the seventh century when beliefs about dragons and other mythical creatures were not uncommon, modern sightings did not occur until the end of the nineteenth century and the monster myth exploded in the mid-twentieth century with the release of a grainy black-and-white photograph supposedly of a long-necked creature breaking the surface of the lake. Subsequent investigations failed to find the monster and despite periodic reports of new sightings, it is now widely believed by scientists to not exist and that the photographs and other supposed evidence consist of either deliberate hoaxes or are due to misidentification of more mundane objects.

There are many sound reasons to think that no such animal could possibly exist. For example, for a species to survive for so long, there have to be many specimens that are able to reproduce over time and yet we have never found even one dead specimen, even though the lake has been sealed off from the ocean for at least 10,000 years. Furthermore, the lake does not seem to have the nutritional resources to support a family of large reptiles over such a long time. In addition, searches using sonar devices and submersibles have failed to detect anything remotely resembling such an animal.

The scientific case against the existence of the monster is overwhelming, and sufficient for scientists to conclude that there is no such thing but this will be convincing only to those who understand how science arrives at definite conclusions about nonexistence (Palci 2016). True believers can still cling to the idea of the monster because there is no way to conclusively prove its non-existence, short of *simultaneously* showing that no part of the lake contains any life form that resembles it. This is an impossible task, since the lake is quite large with a surface area of 56 square kilometers and, with a maximum depth close to 230 meters, contains the largest volume of water of any lake in the British Isles. Even if one were to drain the lake and found nothing, believers could argue that the monster had escaped in the stealth of the night to another lake or even the ocean.

A similar kind of reasoning was adopted in the nineteenth century by believers in "vitalism," that life required the existence of some mysterious vital essence that did not have a material basis.

> No matter what progress was made by biologists to explain life in purely mechanistic and materialistic terms, some aspects always remained unexplained. There were always some corners in which notions of "soul" or the "pure vital force" could find refuge. (Latour and Woolgar 1979, 168)

As long as believers in the existence of some entity can add on auxiliary assumptions at will, they can find a way to counter any argument against their beliefs, similar to the way that even the most bizarre conspiracy theories never die however many individual pieces of purported evidence are refuted.

Believers in the existence of things for which no positive evidence has been provided can and do often make the seemingly reasonable and persuasive argument that since neither existence nor nonexistence have been conclusively proven, the fair-minded person should keep an open mind and allow for the *possibility* of their existence. They argue that the two options listed here, of existence or nonexistence, are not sufficient and need to be expanded to include a third one: It *may* exist even though it has not been shown to be so.

Given how hard one must work to arrive at definite conclusions of existence or nonexistence, one should not be surprised that so many people choose to sink gratefully into the soft cushions provided by this third option. For most people, this kind of ambivalence seems like the most reasonable stance since to take either end of the spectrum (definite existence or definite nonexistence) in the absence of conclusive proof either way smacks of dogmatism and people tend to shy away from taking firm positions that they cannot themselves justify. This kind of agnosticism on the existence of entities that have not been conclusively established is common. The scientist who is convinced of the nonexistence of such things is placed in the position of being labeled the dogmatist in the debate, for not keeping an open mind.

But what exactly is agnosticism? Its roots lie in relation to belief in gods, and the first definition in the Merriam-Webster dictionary reflects that because it says that "agnostic" is the label given to "a person who does not have a definite belief about whether God exists or not." Over time the meaning of the word has been expanded beyond its religious origins and now is used to denote a stance that refuses to commit for or against a particular proposition, as evidenced by the second definition of an agnostic as "a person who does not believe or is unsure of something." Since it is *logically* impossible to prove the nonexistence of *anything*, however absurd (whether it be vampires or ghosts or mermaids or lake-dwelling monsters) except under highly artificially contrived situations, this automatically excludes "no" as the *purely logical* answer to whether anything exists. So if the answer to an existence question isn't a definite "yes," then the only viable alternative seems to be the third option listed that "It *may* exist even though it has not been shown to be so."

But are we condemned to a state of permanent agnosticism about pretty much everything, such as the Loch Ness monster? Is it really the case that one must always allow for the *possibility* of existence of anything that one cannot prove does not exist? Surely not. In purely practical terms, if one believed in the possible existence of every conceivable thing, however bizarre or fanciful, one could not function because one would be constantly terrified by the possibility of monsters lurking under one's bed, friends and family members who may have turned into vampires and seek to drink one's blood, and ghosts popping out at you unexpectedly. Most people do not live such terrified and suspenseful lives, though

I am sure that all of us have members of our family, friends, and acquaintances who live in fear of dark and lonely places and graveyards because of the fears expressed in the traditional Scottish prayer:

> From ghoulies and ghosties
> And long-leggedy beasties
> And things that go bump in the night,
> Good Lord, deliver us!

Others who do not live such fearful lives tacitly adopt rules using an ad hoc approach based on personal whims that tell them that some things either don't exist or that the chances of their existence are so slight as to not be worth worrying about. This enables them to go about their everyday lives without fear, unless they happen to be in a situation that reminds them of the possible existence of such entities (such as visiting cemeteries at night or reading books and watching films about ghosts and vampires) or they live in highly superstitious communities where such beliefs form part of everyday life and conversation. Depending on their situation, most people act as if some things don't exist while leaving open the possibility for others to exist, even though both sets may have, or more precisely lack, the same levels of evidence. So they might say that they don't believe in vampires but do believe in ghosts, even though they would be hard pressed to state exactly how they discriminate between the two cases.

In the absence of any convincing evidence for existence, what distinguishes those things that many people confidently assert don't exist (like vampires, unicorns, and mermaids) from those things (like ghosts, gods, and immaterial souls) whose nonexistence those same people are loath to proclaim? Is there a way to escape from a state of permanent agnosticism, at least in some cases?

It is *scientific logic and reasoning* that saves us from such a "demon-haunted world" (Sagan 1997). In science, we do not shy away from firm statements of nonexistence because we have established methods for doing so because we *need* to. Agnosticism on the possible existence of entities is usually a *temporary* stance pending ongoing investigations but at some point, scientists feel confident in declaring nonexistence and closing the case. This is because there is a difference between saying one is agnostic because of the logical impossibility of proving a negative, and being an agnostic because the evidence is not (as yet anyway) convincing either way. The former leads to a permanent state of agnosticism while the latter is a temporary one.

As an example, for nearly fifty years the currently popular theory of elementary particles postulated the existence of a particle known as the Higgs boson that had, prior to 2012, not been directly detected. The theory of physics that predicted its existence, known as the Standard Model, was so powerful and successful that most particle physicists were convinced that the Higgs did indeed exist because it formed an integral part of its explanatory structure. The theory made predictions based on the assumption that this boson existed and these predictions seemed to be confirmed. This gave scientists confidence that they were on the right track and that direct detection of it would eventually occur. But there were others who hoped that this would not happen because that would signify a breakdown in the Standard Model and provide an incentive for investigation of alternative theories that did not require the existence of the Higgs. New theories are often more exciting to work on than mopping-up operations on old ones because they open up fresh horizons for research. Until the Higgs was actually detected, it was perfectly reasonable to be agnostic on the question of its existence.

The Large Hadron Collider at CERN that was constructed at enormous cost had the detection of this particle as one of its major goals and in 2012 two research teams announced that they had been successful in this search. Once the evidence for the existence of the Higgs was actually established at a 99.99994% confidence level, almost all agnostics about the Higgs shifted out of that camp and into the camp of believers. Note the important point that with deep inferential knowledge *we never get 100% certainty,* and determined skeptics about the existence of the Higgs can still cling to that tiny chance that they are right. There is no way to *prove* them wrong. Indeed in many areas of science, results are published that may have as high a probability as 20% that they were arrived at by chance, giving skeptics much greater leeway to hold on to their beliefs.

If the search for the Higgs had gone on and on being fruitless, would we have been condemned to being agnostic about the particle forever, as some seem to be in the cases of the Loch Ness monster, the Yeti, Bigfoot, vampires, and ghosts? Would it be impossible to conclude that the Higgs did *not* exist? No. A state of permanent doubt in such situations is unwarranted. Using the logic and methods of science, such states of agnosticism can be terminated and conclusions of nonexistence arrived at. This has happened in the past in the cases of the vital essence, aether, N-rays, polywater, and phlogiston, each of which laid strong claims to existence at various times but are now firmly believed to not exist and do not enter into any of the currently prevalent scientific theories.

How was their nonexistence established by scientists in the absence of proof? A particularly good example involves the case of the aether. This was believed to be a material substance that permeated all of space and whose existence was deemed to be necessary in order to explain the propagation of light and other electromagnetic waves, similar to the way the Higgs particle was postulated to explain the workings of the Standard Model. As more and more experiments failed to directly detect the aether, its properties had to be refined and modified to explain away the negative results, not unlike the case with vampires, ghosts, and the like. Even though the theory of the aether became quite convoluted as a result of these additional suppositions, it was still thought to exist because *it was necessary as an explanatory concept* for understanding the behavior of light. Belief in the necessity of the aether was so strong that James Clerk Maxwell (1831–1879), whose set of equations forms the basis for electromagnetism to this day, reportedly thought that the aether was better confirmed than any other theoretical entity in natural philosophy (Laudan 1984, 114). The eminent physicist William Thomson (1824–1907), better known by his later title of Lord Kelvin, also believed that the aether was real and substantial, and Heinrich Hertz's experiments on radio waves in the period 1886 to 1888 were seen as confirming its existence (Chalmers 1990, 64).

When Einstein came along with his theory of relativity, he did not, contrary to popular belief, *prove* that there was no aether. He could not possibly do so for all the reasons we have discussed here. When it comes to the aether, as John Bell says, "The facts of physics do not oblige us to accept one philosophy rather than the other" and it is still possible to use the aether as a heuristic tool for calculations (Bell 1976, 77). What Einstein did with his alternative theory was make the aether *irrelevant* and *superfluous*. As a consequence, scientists now comfortably assert that the aether does not exist even though we have not proved it, and despite having once firmly thought that it did exist.

The case of the aether illustrates a very important point: In science, when there is no credible *positive* evidence in favor of the existence of something *and* it also *becomes unnecessary as an explanatory concept*, it can be confidently asserted to *not* exist (Singham 2011). For this reason, scientists are no longer agnostic on the question of the aether's existence. For all intents and purposes, the aether does not exist. It has ceased to be. Using this same line of reasoning, I feel confident in saying that the Loch Ness monster and the Yeti and Bigfoot and vampires and ghosts do not exist either and I live my life without any fear of them suddenly popping up.

The basic problem with the argument that if something cannot be proven to not exist then one must leave open the possibility of existence can be traced to treating the two initial statements of existence and nonexistence as being on an equal footing and then weighing the evidence and arguments for each and seeing which comes out on top. As demonstrated here and will be discussed in more detail in chapter 14, *that is not how scientific logic and reasoning works*. When it comes to existence statements for any entity, the *default* position that is presumed to be true is that of *nonexistence*. The burden of proof is on the person advocating for existence to produce a *preponderance of evidence* in favor of that position. In the absence of that level of evidence for existence, and if an entity is unnecessary as an explanatory concept (in that its absence does not cause any logical or theoretical problems) *nonexistence is taken to be the true situation*. If pressed, a scientist will always leave open the *theoretical* possibility that new evidence might emerge that proves her to be wrong but in practice nonexistence is taken for granted. This is how agnosticism about the existence of an entity is terminated.

But this is just one of the ways in which scientists arrive at firm conclusions of nonexistence in the absence of logical proofs. The rest of the book will develop the idea of scientific logic and reasoning in many other ways that enable us to greatly expand our ability to make definitive statements about things that lie outside our range of direct experience and be confident about not only the existence and nonexistence of entities but also about the theories and laws that we depend upon all the time in order to be able to function.

5

The power of scientific theories and the problem of induction

[Despite the counterintuitive nature of many scientific conclusions, it is because science works so well that people accept them. While the problem of induction prevents us from generalizing from a few instances and predicting the future purely by what we have observed in the past, it is the belief that the laws and theories of science are the best that we have and getting better with time that gives us confidence in their predictions. The importance of having a better understanding of the way that the scientific community uses the words "law," "theory," "hypothesis," and "fact" is emphasized.]

Here's an exercise. For each of the following statements, which side of the "OR" is true?

1. The Earth orbits the Sun OR the Sun orbits the Earth.
2. The Earth rotates about its own axis OR the universe rotates around the Earth.
3. The Earth's continents move very gradually OR they are fixed.
4. The universe is expanding OR the universe is static or contracting.

Anyone who has listened to their teachers during their school science classes or pays even cursory attention to the reporting of news on the world around them would agree that the first of each pair of options is the right one: that the Earth orbits the Sun, it rotates about its own axis, the continents move, and the universe is expanding. And I would expect that most of them would be sure that they are right.

But why are they so certain about all those things? What gives them that confidence when their everyday experience argues so strongly for the opposite? That experience, after all, is why the alternative set of statements were believed to be true for millennia, and that the continents are fixed and the universe is static were believed even as late as the mid-twentieth century.

Many people might say that these things are true because *science has proven them to be true*, even if they themselves cannot lay out exactly the chain of evidence and reasoning that led to those conclusions. That is because these counterintuitive conclusions were only arrived at after hard work done by generations of scientists who performed increasingly delicate and difficult experiments and used observations, careful reasoning, mathematical techniques, and calculations that are not within the skill range of the general public. This knowledge has now become part of such a solid consensus within the scientific community that they are embedded in the science textbooks that are used to teach students. These counterintuitive beliefs are now firmly backed by the authority of science, and science is believed to have earned that high level of credibility.

We place our faith in science because it has been successful to such an extraordinary degree. Perhaps nothing shows this more clearly than the modern attitude toward airplane flight. Many thousands of people get on planes every day and think nothing of the fact that they will be soon flying in a hollow metal tube at high speeds over vast distances at many thousands of meters above the ground, an incredible thing when you pause to think about it. The biggest worry for most people is not that they might hurtle to their deaths but that their luggage may get lost. Similarly, we expect our cars to start at the turn of a key or the push of a button and our GPS systems to take us to the correct destination. We have got so used to the reliability of smartphones that we expect them to work flawlessly even when talking to people on the other side of the world and we get annoyed when a call is dropped. In the field of medicine, we have developed drugs and vaccines and methods of treatment that are so powerful that we can successfully cure many illnesses that used to be fatal, and have eradicated or almost eradicated diseases such as smallpox and polio that used to ravage many parts of the world. We have had people go to the Moon and return, sent mobile vehicles to roam the surface of Mars, and there are probes that have even left the solar system on their way into deep space. All this is due to advances in science.

Science works. Even though it is by no means perfect or infallible, knowledge that carries with it the imprimatur of science is considered to be more reliable than that provided by any of the alternatives. Even when things fail, we tend to blame it on factors like machine or human error rather than due to the failure of science. But what enables us to say this? Why *is* science so successful? Surely it must be because we believe, as Aristotle did a long time ago, that science has found a way of distinguishing between true knowledge and mere opinion.

Of course, this raises the immediate problem that if science creates or discovers true knowledge, how can it be that scientific theories that were once thought to be true were later deemed to be false? One possibility is that the scientists of the past were simply not as good as the ones we have now and made errors. Another is that our modern technology enables us to do more careful and precise studies. A third is that over time we have managed to recognize and eliminate almost all the wrong ideas. The fourth is that we are just lucky and happen to live at a time when scientists have finally figured out how the world actually works, at least mostly. But if one is not as hubristic as to think that we have somehow finally discovered the true theories, a more modest stance is to see scientific theories as steadily *getting better*. Even if we cannot say for sure that what we have now is the last word or the ultimate truth, perhaps we can at least say that current theories are better than those of the past, more true if you will. In this view, science is steadily progressing onward and upward toward the truth and will someday get there, even if not in our own lifetimes or those of our children and grandchildren.

Part of the reason for our confidence in science is because its empirical success is indisputable. Science does indeed work exceedingly well and yet the power of science is not due purely to its empirical success. While that plays an undoubtedly major role in creating our confidence in science (it is because planes crash so rarely these days that we have little doubt concerning the validity of the scientific theories that are the basis of their design, manufacture, and operation), scholars say that it cannot be the sole basis because we immediately run into the well-known *problem of induction* that was raised by philosopher David Hume, that states that there is no justification for generalizing from a few specific instances to a universal statement about all possible instances. One form of this is that just because all the members of a class of objects that we happen to be aware of share a particular property, that does not justify believing that *all* the members of the class must do so. A famous example was the belief that all swans are white, something that was strongly believed until people discovered the existence of much rarer black swans.

A related form of the problem of induction is that just because something has happened many, many times in the past without exception, that *by itself,* is no reason to think that it will do so in the future, or even the very next time. An example of this is that the Sun will rise tomorrow. The Sun has risen in the east for every day of our waking lives, using the common formulation and ignoring the reality that the Sun does not "rise" and "set" but that it is the Earth's rotation about its own axis that creates that illusion. It has done so for the 4.5 billion years that the Earth has been in existence. This seems to be as certain a fact as anything, as evidenced by preambles such as "As surely as the Sun rises" being commonly used when predicting something that the speaker is certain will happen, found even in the Bible (Hosea 6:3).

But the problem of induction says that there is no *logical* reason to think that the Sun will rise again tomorrow just because it has always done so in the past. In fact, just like some people believed that all swans were white because black swans were only found in a few parts of the world that they had not yet visited, so it is with daily sunrise. The cessation of sunrise actually does occur *every year* in parts of the world where few live and thus the rest of us do not experience it. If one lives within the Arctic or Antarctic circles, there will come a time each year when the Sun sets and does not rise again for a few months. We can explain that away as being due to the tilting of the Earth's axis, that the Sun still does "rise" and "set" in the two polar regions but that it occurs below the horizon so that we cannot see it from that part of the planet.

But there are many things that repeat themselves without fail day after day but then stop without warning. Our lives for example. Each night we lose consciousness for a period by falling asleep and later regain it by awakening. We have done so every day of our lives. But yet there will come a day when there will be no waking up. In fact, there is no guarantee that this will not happen the very next time we go to sleep. This is not a cheery thought but a reminder of the cautionary words of the commercials for investment firms that are required to warn us that past performance is no indicator of future success. But while dying is regrettable and many people seek some facsimile of immortality through wealth, fame, monuments, and achievements (though as Woody Allen once said, "I don't want to achieve immortality through my work. I want to achieve it through not dying"), we are not shocked by the idea that there will be this one exception to an unbroken string of successes of waking up after sleep.

And yet whether we wake up to see it or not, the idea that the Sun will not rise tomorrow morning is unthinkable. We have an unshakeable confidence that it will do so because *science predicts that it will*. So clearly when it comes to the sunrise, we have a level of confidence that goes well beyond the empirical facts of unbroken past success and enables us to ignore the problem of induction. While empirical success is a necessary condition for such confidence, it is our belief in the rightness of *scientific laws and theories* that enables us to have this high level of it. The rotation of the Earth about its own axis is governed by the laws of physics and those laws *predict* that the Earth will rotate on its axis for all time (though very gradually slowing down due to tidal friction), unless we are the victims of a cataclysmic cosmic event that destroys it, such as when our Sun burns out and becomes a Red Giant that swallows up the Earth or if our galaxy at some point collides with another galaxy and the Milky Way gets ripped apart or some other catastrophe. For the Earth to stop rotating for no discernible reason would be to *violate the laws of science*, and so we expect that it will never happen. Biological theories and laws, on the other hand, offer no guarantees of immortality and indeed our eventual deaths are wholly consistent with them.

Our unshakeable confidence that the sun will rise tomorrow morning is due to our belief that the *laws of physics are true and unchanging* and thus the predictions that ensue from them are so reliable that we can rely on them. Similarly, our confidence that the airplanes we get on will not crash is due to our belief in the reliability and consistency of, among other things, the laws of hydrodynamics, thermodynamics, electrodynamics, and mechanics. Our belief that we can call our friends across the globe at any time is due to our confidence in the laws of electromagnetism, and so on. Even when failures occur, such as planes crashing and calls getting dropped, we look for local causes for the problem and never question as to whether the laws themselves were violated.

In short, what many people may not realize is that when they make casual predictions about the future, such as making plans for the coming winter and the subsequent spring, what they are really doing is expressing *deep and unshakeable confidence in the theories of science.* We tend to treat the theories and laws of science that provide us with all this success as so reliable that they must be true, now and for all time. But what makes us think that?

At this point it is necessary to make a slight digression into vocabulary. The words used in science can often be confusing and misleading for the general public. This is due to two reasons. Scientists often co-opt everyday words for use in science but restrict their meanings so that their technical usage is different from their everyday one. Conversely, words that originate in science and had limited technical meanings at one time (such as "velocity," for example) often transition to everyday use and become looser in their application. Since language plays an important role in understanding the state of knowledge within science, getting agreement on what key words mean is important if we are to have a meaningful dialogue between scientists and nonscientists, and the difference in usage is the source of many of the confusions and misconceptions that nonscientists have about science. Alert readers will already have noticed that in the preceding remarks, I have used the words "theory" and "law" almost interchangeably when describing science and for those who may be bothered by this I need to clarify as to why I am doing so, since these and two other words, "hypothesis" and "fact," all occur repeatedly in any discussion of science and are used frequently in describing its investigative processes.

Suppose we think of the different kinds of knowledge represented by these four words as occupying a hierarchy, ranging from the knowledge we are least confident about to that of which we are most certain. Everyone, scientist and nonscientist alike, will agree that a hypothesis is a sort of educated guess, and occupies the lowest rung of the hierarchy and is the least certain. They will also agree that at the top of this hierarchy lies a fact that forms the most certain kind of knowledge, consisting of something that we are absolutely sure of, such as a quantity that has been either directly observed or experimentally measured. Theories and laws lie somewhere in between, but is there a hierarchy to them too?

Theories play an essential role in constructing scientific knowledge. While it is asserted that science is an empirical enterprise subject to experimental tests, we do not think of scientific knowledge as consisting of only empirical knowledge and data. We want our knowledge to be useful, to have *explanatory power* and also enable us to *make predictions* about the future, and it is scientific theories and laws that enable us to do so and provide context and meaning to experimental results. It is important to realize that *it is from theories and laws that science derives its immense power.*

But the words "theory" and "law" are used in very different senses within the scientific community as compared to those outside it, and this influences their positions within the hierarchy. Nonscientists often use the word "theory" to signify something that it is unproven (close in meaning to the word "hypothesis," but a little more elaborate) while a law is seen as being a theory that has been *proven to be true* and is thus more like a fact. When used in popular discourse, the word "theory" is sometimes prefaced by the word "mere" or "only," thus emphasizing its lowly status. In fictional crime literature, the stereotypical unimaginative police officer who sneeringly responds to another's idea by saying "That's a nice theory but where are the facts?" is reflecting this common view. As a result, for the general public a law carries much more weight than a "mere theory" and thus occupies a higher status in the hierarchy of knowledge. So in the popular mind we have a knowledge hierarchy in which hypotheses lie at the bottom with theories a little above it, while facts lie at the top with laws just slightly below it.

This popular understanding of the meaning of the word "theory" has been exploited to form the basis of attacks on science, such as the teaching of the theory of evolution in schools. Critics of evolution frequently argue that science is supposed to concern itself only with the teaching of things that are certain and thus should confine itself to experimental results (the "facts") and the *laws* of science, and not propagate mere hypotheses or theories. They then proceed to make the charge that since scientists themselves refer to the prime mechanism of evolution as the "*theory* of evolution by natural selection," they are tacitly acknowledging it to be unproven. Hence, these critics assert, evolution should either be omitted from the curriculum altogether or be taught along with other things that are believed by many but cannot be proven to be laws or facts, such as biblical or other theories of creationism.

Nonscientists are sometimes a little surprised when I tell them that in science there is no hierarchical distinction between theory and law and that there exists no scientific body whose role it is to evaluate theories to see if they have passed some test of truth and can thus be elevated to the status of a law, or to demote a law to a theory or even consign it to the scientific dustbin if it is found to have subsequently failed the test of truth. It seems incredible to them that scientists attach no special significance to whether something is called a theory or a law. This is because scientists understand the words "theory" and "law" quite differently from nonscientists. Both words are used by them to signify intellectual constructs that *infer the existence of patterns* on what would otherwise be merely collections of unconnected facts. There is no value judgment about truth that can be inferred based solely on these labels. As an example, Einstein's special theory of relativity is on far firmer scientific ground than Newton's laws of motion, in the sense that there are no known serious violations of the former (that is labeled a theory) while the latter (although labeled as laws) are known to fail under certain conditions.

Whether something gets called a law or a theory (or sometimes a "principle") is an accident of scientific history and nothing more. *To the scientist, they do not signify different points in the knowledge hierarchy.* A law, just like a theory, is arrived at *inductively* based on experimental or observational data that suggest a pattern that is assumed to hold more generally. Both theories and laws are understood by scientists to be products of human imagination and creativity, and their purpose and function is to provide coherent and meaningful explanations of observed phenomena and to make predictions about things not yet seen. Theories and laws do not occupy different points on the knowledge hierarchy.

One useful, though not rigid, distinction that can be (and is) made between the words "theory" and "law" is to reserve law to describe an inductive generalization suggested by selected observations and that often consists of a single statement or equation, while a theory is a more detailed *conceptual scheme* or explanatory framework that often postulates some kind of mechanism that can be used to interpret experimental data. For example, Boyle's law is a statement that is a generalization based on selected observations on the pressure and volume of gases. It deals with the macroscopic properties of gases and can be expressed by a single formula "pV=constant" that can be expressed in words as "the pressure exerted by a fixed quantity of gas multiplied by its volume will retain the same value even as its pressure and volume individually change." This law enables one to *predict* how the volume of a gas would change if one changes the pressure on it and vice versa. But it does not tell us *why* gases behave this way.

The *kinetic theory of gases* however is not a single statement at all but is a model that makes certain assumptions about *individual* gas molecules and their interactions that seeks to explain the behavior of gases at a microscopic level. The kinetic theory can be used to explain how Boyle's law, among other macroscopic properties of gases, comes about. It also explains why Boyle's law only strictly applies to what we refer to as "ideal" gases and is only approximately true for real gases. Thus both Boyle's *law* and the kinetic *theory* of gases play *complementary* roles in our understanding of gas behavior. Neither one is truer than the other.

Practicing scientists would, however, agree that theories and laws are different from scientific facts, which are believed to be experimental observations and measurements made under rigorous, and usually repeatable, conditions. It is the interplay between theory (or law) and experimental fact that is the key to understanding the traditional model of how scientific knowledge progresses. The scientist does this by observing the world or by making measurements on specific features of the world and then trying to identify patterns or systematic features from those observations. If a pattern seems to exist, the scientist *inductively* postulates a tentative hypothesis about why it might be so and if the hypothesis turns out to have general utility and meets with some success, it may be elaborated into a broader explanatory framework that enables scientists to investigate it further, and this broader construct is sometimes called a theory or a law. Thus theories and laws are acts of human imagination that are arrived at inductively, going from a few, specific, concrete instances to a general pattern that explains, governs, and predicts behavior on a wider scale. Theories and laws are thus subject to the problem of induction and the reasons why we believe so strongly in some of them despite that limitation will be discussed later.

This identification of patterns and ascribing reasons for them is not limited to scientists but is something that *everyone does all the time*. So for example, one can imagine early humans observing the daily sunrise and inferring that there must be a reason for its inevitability, perhaps proposing the hypothesis that there was a Sun god in control of its motion who liked things to proceed in a regular fashion and made sure that the process repeated itself without fail.

So contrary to the view of critics, science is not about facts but is all about theories, with facts serving as sources of evidence. Far from being "mere," *theories are the most powerful tools we have* in science because they enable us to understand phenomena and predict behaviors. Hence the idea that there should be some sort of test to determine if a theory is true seems perfectly reasonable. After all, if scientists believe that their goal is to seek the truth about the world, and if theories are the means by which we interpret the world, then scientists should have a mechanism to judge whether they are true or not. Otherwise how can we be confident that science is actually progressing, continuously rejecting older false theories while accepting newer theories? This is why it is so important for the scientific community to have agreed-upon yardsticks for comparing the merits of competing theories. The way scientists make those judgments forms the core of this book.

What about facts? Aren't they the only things that we can confidently assert to be true? Actually no, not always. Those facts that are accessible via direct observation, such as the existence and characteristics of the desk at which I am sitting, we may confidently assert to be true. But our physical senses are crude instruments that are only capable of observing things within a highly limited range. Most of the facts that we believe in, especially in science, are of a *deeply inferential* nature and *become facts only within the context of the theories used to generate them*. In 1921, on the occasion of his fiftieth year as the editor of the *Manchester Guardian* newspaper, C. P. Scott wrote in an essay that "comment is free, but facts are sacred," a sentiment that is now widely repeated. But the reality is that when it comes to knowledge that is arrived at using the deep inferential methods of science, facts are not only not sacred, *they are not even facts*, at least as far as being considered unchanging objective realities.

Take for example the existence of electrons. They are outside the range of direct sensory experience. Nobody has seen, touched, smelled, heard, or tasted one. So why are we so sure that electrons exist? The reason we firmly believe they exist is because an entity with a certain set of properties to which we attach the label of electron has been *postulated* as an *explanatory mechanism* in order to explain a wide array of observations, and it has been incorporated into an extensive theoretical framework. In investigating that framework, a vast number of experimental observations have been so consistent with the assumed existence of electrons that we now have little doubt that it exists. But its existence is a fact only within the context of theories that predict, allow for, and depend upon it for their validity. As a result, it may be reasonably argued that an electron is a theoretical construct, or a metaphor. But even so, that does not mean its existence is not a fact because we have every reason to believe in its existence as long as the theories that depend upon it are considered to be true. We thus see that theories, hypotheses, laws, and facts form an interconnected web of concepts, each playing an essential role in the construction of knowledge.

In the second part of the book, I will examine in some detail a case study of the search to pin down a number for the age of the Earth that shows this interconnectedness and how this particular scientific fact has changed over time as the theories used to answer this question have changed. I have chosen to look at this particular question out of many possible ones because it took a long time to reach consensus on the answer, the theories that are required to arrive at that consensus verdict span a wide range of scientific disciplines, and the science needed to understand the process is not too difficult or esoteric.

PART TWO

CASE STUDY OF THE AGE
OF THE EARTH

6

What the age of the Earth reveals about how science progresses

[A detailed case study of the search for the answer to the question of how old the Earth is serves as a paradigm for how scientific "facts" need not be unchangeable. The age has oscillated wildly before settling on the currently accepted value of 4.54 billion years, with religion, politics, and other nonscience factors influencing the search along the way. The final consensus involves a complex interplay of theories from geology, biology, physics, chemistry, and paleontology.]

How old is the Earth?

A case study of how the scientific community investigated what seems like a straightforward and basic question will reveal many features that illustrate the workings of science. While posing the question may be simple, arriving at a consensus judgment as to the answer is not. Other case studies that exist in the literature, such as how the Copernican model of the solar system triumphed over the Ptolemaic one (Kuhn 1957) or the Copenhagen interpretation of quantum mechanics became the orthodox view over alternatives (Beller 1999), could have served this purpose just as well because the resolution of major scientific questions share many common features. I chose the age of the Earth because the question of fact being addressed is easy to state and the knowledge required to understand how the answer evolved is less technical and esoteric and thus better serves my purposes of illustrating to a general audience the interplay in science of theories, laws, facts, and hypotheses.

The age of the Earth should be a simple, unchanging fact but this has not been the case for many millennia, with various values being believed at various times depending on the theories being used to answer the question. The changing answers correlate with the evolution of theories in such widely different fields as geology, physics, biology, chemistry, and paleontology. The process was also shaped by technology and the cultural, political, religious, and other nonscience factors that created the climate in which the various theories were developed. As a result, over time the answer to the question of how old the Earth is has oscillated from infinitely old to really young (thousands of years) to quite old (hundreds of millions of years) to somewhat younger again (tens of millions of years) to the currently accepted value of 4.54 billion years, though some people are still stuck in the second phase and believe that it is only a few thousand years old.

(a) Early models of the age of the Earth

While there are of course no written records to substantiate what was believed in very early human history, one can easily construct plausible scenarios about how early humans might have arrived at different assessments of the Earth's age based on what they saw around them. Long before they could come up with numerical estimates, one can imagine that people felt that the Earth was so large and gave the impression of such solidity and permanence that it must have been around forever. Aristotle, for one, had argued for an eternal, uncreated universe (Grant 2004). On the other hand, one could also envisage early people having a cyclic model, as suggested by the regularity of the seasons in which summer invariably follows spring, then fall, then winter, and then spring again. Similarly, night follows day, all without fail. The seductive principle of induction that things that have always happened will continue to happen also works backward and this could have suggested to early humans that these cycles had been going on forever and that the Earth, and indeed the entire universe, also went through cycles of creation, destruction, and rebirth and had undergone an infinite number of such reincarnations.

One can find both models, the eternal and the cyclic, in various mythologies and creation stories and, in the absence of any actual data, the answers to these questions were provided largely by mystics according to the needs of their particular cosmologies. The ancient Egyptians considered the age of the Earth to be vast but indeterminate, while many Greeks and eastern cultures adopted cyclic models (Burchfield 1975, 3). Hindu tradition has a quite specific cyclic cosmology in which the universe is believed to endure for about 4.3 billion years (1 Brahma day) and then is destroyed by fire and water and then re-created again (Jackson 2006, 6). Though the Buddha himself refrained from articulating any specific theory about the origin of the universe, the cyclic model might well have appealed to Buddhists who believe in reincarnation.

The Bible imposed a different set of constraints in those societies in the west where it was the dominant religious influence. The book of Genesis with its vivid, detailed account of the creation of the universe cemented the idea that the universe had a specific beginning. The story of the Garden of Eden and the fall from grace of Adam and Eve that required Jesus to save humanity from that original sin ruled out cyclical models, since that would have required multiple falls from grace and multiple arrivals of a savior. The Genesis story thus requires people to believe in the idea that the Earth had a unique emergence with a definite beginning and thus a finite age. Thus the age became a well-defined problem whose answer should be discoverable and, as we will see, such problems have always had great appeal for the scientifically minded and form the basis of much scientific research. What the early problem-solvers lacked, however, were any secular theories or data to help them uncover the answer. As a result, until around the mid-eighteenth century when geology started to become an established and independent field of scientific study, the question of the age of the Earth was largely left to religious scholars to address and several of them arrived at answers based on what they could glean from their religious texts and related sources.

(b) Young Earth theories

Christians in particular had a deep interest in addressing the question of the age of the Earth because the second coming of Jesus was an eagerly anticipated event. Initial expectations were that Jesus would come again during the lifetime of his disciples or shortly thereafter, but as time went by with no arrival, those predictions were abandoned and people settled in for a longer wait. They started looking in the Bible more closely for hints of when the second coming would occur, a practice that some indulge in even today with websites devoted to finding markers that indicate how imminent that day might be.

Irish bishop James Ussher (1581–1656) set about doing just that calculation. The impetus for his work was a belief that Jesus's return would also result in the end of the world. A common belief among Christians was that the world would last 6,000 years, this being an interpretation of the six days of creation in Genesis combined with the statement that "For a thousand years in your sight are like a day that has just gone by, or like a watch in the night" (Psalms 90: 4) and the New Testament statement that "With the Lord a day is like a thousand years, and a thousand years are like a day" (2 Peter 3:8). The Genesis story was taken both literally, as an actual depiction of historical events, and allegorically, with the creation of humans on the seventh day being taken to indicate when the second coming of Jesus would occur. It was believed that we were living in the seventh and last age. Thus the age of the Earth was an important piece of information because if you knew how long the Earth was predicted to last, then fixing the date of creation would enable you to calculate when it would end with Jesus's return. You could then prepare yourself for that grand event.

Ussher used as his sources all the ancient texts available to him in Latin, Greek, and Hebrew, with the Bible being considered the most reliable (Rudwick 2014, 12, 14). Within that framework he did a remarkably precise calculation, declaring in 1650 c.e. that the Earth had been created in 4004 b.c.e. Ussher arrived at his creation date by first fixing the date of the earliest event in the Bible that could be corroborated with other historical sources and then working backward. The death of the Babylonian king Nebuchadnezzar in 562 b.c.e. (a date which is known independently of the Bible) is reported in the Bible to have coincided with the thirty-seventh year of exile of king Jehoiachin of Judah (2 Kings 25:27), although the dates are sometimes off by a few years because of the differences in the calendars in use at that time.

From that fixed reference point Ussher worked backward, first by adding up the years that the successive kings ruled the divided kingdoms, the southern one of Judah and the northern one of Israel, and when that vein of information ran dry because the kingdoms did not exist before a certain time, going further back using the famous "begats" in the Bible, which give a detailed genealogy that goes back to Adam. For example, Genesis chapter 5 gives the chronology from Adam to Noah, and then after a long story about Noah and the Great Flood, chapter 11 continues the genealogy from Noah to Abraham. Curiously, the Bible conveniently provides the age at which each man became the father of his eldest son, all the way down the male line. It is a piece of trivia that one might not expect scribes to bother to record, but that information was invaluable for Ussher's method of calculation.

Ussher initially arrived at a date of 4000 B.C.E. for the year of creation. Thus according to his chronology, Jesus was born 4,000 years *after* the creation and since the world was expected to last for 6,000 years, it would end in 2000 C.E., which provided a pleasing rounding of numbers. However, those dates had to be adjusted because the creators of calendars had made a mistake in setting the marker for when the Common Era (C.E.) began. When corrected, it turned out that King Herod had begun his reign in 37 B.C.E. and died in 4 B.C.E., and Jesus had to have been born some time during that period because the New Testament says Jesus was born during Herod's reign. Ussher fixed the year of Jesus's birth as 4 B.C.E. and this required the shifting of all the dates by four years, moving the year of creation to 4004 B.C.E. Once he had that anchor date, he could use the Bible to also pin the dates of other major mythical biblical events that have no corroborating evidence, such as Noah's Flood (2348 B.C.E.), God's call to Abraham (1921 B.C.E.), and the exodus from Egypt (1491 B.C.E.). It should be noted that although we now think that 6,000 years is a ridiculously low figure for the age of the Earth, at that time it was considered an enormous span of time, long enough to go well beyond all recorded history, though this required ignoring ancient Egyptian, Chinese, and Babylonian texts that suggested that human history might extend back tens of millennia (Rudwick 2014, 27).

It should be noted that Ussher was not the only, or even the first, person to do such calculations, just that his are the best known. Other people had done similar calculations and arrived at similar results, such as Joseph Scaliger (1540–1609), who fixed the date as 3949 B.C.E. (Rudwick 2014, 14); Theophilus, Patriarch of Antioch, (~115–183 C.E.), who fixed the date of creation at 5529 B.C.E.; Julius Africanus (~160–240 C.E.), who said it was 5500 B.C.E. (Burchfield 1975, 4); and Rabbi Jose ben Halafta (~160 C.E.), who said that Adam was created in 3760 B.C.E. The Bishop of Worcester William Lloyd (1627–1717) also got 4004 B.C.E.

(Jackson 2006, 30). It is not well known that the great scientist Isaac Newton (1643–1727) along with others also used biblical chronology and got results similar to Ussher's, to within about fifty years, depending on their choice of source material of text, translations, and commentaries, of which there were many.

Ussher's calculation likely became the best known one and part of Christian folklore because John Fell (1625–1686), Bishop of Oxford who also served as the controller of Oxford University Press, initiated an annotated edition of the Bible that included Ussher's dates for the major events in the margins, thereby conferring on it strong credibility, though the first editions did not specifically cite Ussher as the source (Jackson 2006, 30). Those dates remained in the margins until 1885, thus cementing those dates in the minds of most of the English-speaking world. So at least in those parts of the world where Christianity was dominant, from 1650 onward there was a fairly broad consensus that we were getting to within a few hundred years of the date of Jesus's return because of the 6,000-year-old limit for the Earth's existence.

(c) The impact of young Earth beliefs on geology

The acceptance of Ussher's calculation for the age of the Earth had implications outside of religion, in particular for the embryonic field of geology. Early geologists were struggling to understand the origins of the major features of the Earth such as the existence of high mountains and steep canyons. The discovery of seashell fossils on the tops of mountains suggested to them that at one time the tops of mountains must have been below sea level and that either sea levels had once risen above them or that the mountains had got pushed up. Nicolaus Steno (1638–1687) and Robert Hooke (1635–1703) also found evidence for layers of geological strata that suggested that various sequential processes were at work in the creation of the Earth's crust (Cutler 2003).

Steno and Hooke did not try to use that information to fix the date of the Earth because it was assumed that that question had already been answered. Their focus instead was on how these major geological features could be consistent with Ussher's calculation for the age for the Earth. Having only a few thousand years at their disposal, scientists of that period were led to the idea that mountains and valleys and fossils and sedimentary rocks had to be caused by sudden cataclysmic events such as earthquakes and floods, including the Great Flood of Noah. This model came to be labeled *catastrophism*, that the Earth's features were shaped by one major catastrophe after another that enabled major geological features to emerge relatively quickly. Rene Descartes (1596–1650) and Gottfried Wilhelm Leibniz (1646–1716) were names closely associated with this idea (Burchfield 1975, 5).

But Ussher's work coincided with the dawn of the Age of Enlightenment and that brought with it the beginning of a separation of scholarly thinking from religious dogma, enabling scientists to speculate more freely and broadly about all matters, including the age of the Earth. As the desire and need for conformity with biblical estimates weakened, scientists started devising alternative theories for the formation of the Earth and the universe that were not explicitly linked to biblical stories. Immanuel Kant (1724–1793) and Pierre Laplace (1749–1847) created a new model of the universe called the nebular hypothesis that used Newton's laws of mechanics and his theory of gravitational attraction to explain the formation and evolution of the solar system. This model said that stars and planets such as the Earth originated as clouds of gases that coalesced under gravity. In the process, their initial kinetic and gravitational energies were transformed into heat energy that resulted in, for the planets, initially molten bodies that over time had their heat dissipate to give us the relatively cool Earth (at least on its surface) with a solid crust that we now have.

Georges-Louis Leclerc, Comte de Buffon (1707–1788), was one of the first to try to pin an actual number to the age of the Earth using only scientific theories and data. He used estimates of the initial internal heat of the Earth and its rate of cooling to arrive at a value of about 75,000 years, a result that he published in 1778 (Jackson 2006, 117). Although that number is wildly off the mark by modern standards, we must remember that he was working before the field of modern thermodynamics and its associated laws governing heat flows had been formulated, and at a time when thermometers were just coming into being. In order to estimate the parameters involved in cooling, Buffon did experiments involving the cooling of metal spheres that used the extremely crude method of actually touching objects with his hands to judge when two objects were at the same temperature. He was aware that his methods were crude and suspected that the actual age of the Earth could be much greater, possibly up to ten million years (Rudwick 2014, 66).

Despite these understandable shortcomings, Buffon's result was a significant development in two respects: it used purely secular scientific reasoning to arrive at an age for the Earth; and the age he reported completely broke with a Bible-based chronology, going well over the roughly 6,000 years that people believed the Bible required. This caused the theologians at the Sorbonne to complain bitterly. This would not have bothered Buffon too much since he was an influential figure and thus immune from the dangers that ordinary heretics might face, though he was willing to make some conciliatory remarks in later editions of his book. Incidentally, Buffon had earlier published the first of his three-volume *Histoire Naturelle* in 1749 which, over a hundred years before Darwin's *Origins* book was published, suggested that the descendants of ancestral organisms, aided by migration and geographic separation, could diverge thus leading to the creation of new species (Henig 2000, 97–98). He also made explicit the increasingly widespread idea that human beings appeared at a very late stage in Earth's history, contradicting the biblical Genesis story that human history and the Earth's history covered the same time span apart from the first five days before the creation of humans. This too had angered the Sorbonne theologians, suggesting that ruffling theological feathers with heterodox views did not unduly worry him.

This new freedom of thought stimulated interest in those areas of knowledge that we now label as geology and paleontology as scientists started to investigate the origins of the Earth and its fossils without the stringent constraints of biblical chronology. As mentioned before, people like Steno and Hooke had earlier observed the presence of seashells and other fossils on mountain tops and patterns in the layers of rock strata, and had used that information to create theories of geological formation in which rock layers were formed by sedimentation, with newer layers of rock settling on top of older ones. But they had not used this insight to actually try to date the Earth, because the biblical ages were the accepted beliefs in their time.

But now their early work formed part of the basis of the new sciences of geology and paleontology, combining the theory of *slow* sedimental formation with the ordering of fossils in the layers of rocks in which they were found. The clear pattern of evolution that emerged in the rock strata (with simpler fossils being found in the layers lower down and more complex ones in layers higher up) led paleontologists such as Georges Cuvier (1769–1832) and Jean-Baptiste Lamarck (1744–1829) to suggest that the process of geological formation must have been quite slow, and required far more time than the Genesis story allowed, even though these early paleontologists were Christians. Steno, for example, was born into a Lutheran family and trained as an anatomist and geologist but later converted to Catholicism, became a monk, and gave up his scientific work in favor of theological studies, and Cuvier was religiously orthodox. But they all felt that the Bible should not be the source of empirical data for investigating the age of the Earth and that the evidence supplied by the Earth itself could reveal its origins. They felt that the truths revealed by the "book of nature" should be *complementary* to the truths revealed by the "book of God" and thus were untroubled by the possibility of any contradiction emerging.

A major development occurred when James Hutton (1726–1797) published a paper in 1785 that put into print ideas that had been circulating widely at that time that argued that catastrophes and great floods were not necessary to explain the features of the Earth; that they could have been caused entirely by the *slow and steady accumulation of small changes* (Rudwick 2014, 68). In 1788 he argued in another paper that not only was the Earth infinitely old, it would also last forever, saying: "The result, therefore, of our present enquiry is that we find no vestige of a beginning—no prospect of an end" (Jackson 2006, 92).

These events were markers of the decline of catastrophism and the birth of the model known as *uniformitarianism*, which was the label attached to the idea that major changes could and did occur because of the steady accumulation of infinitesimal ones, and the forces that were shaping the geological features at present were the same as they had always been. It was clear to this school of thinkers that adopting this perspective meant that the Earth had existed for a vastly longer period of time than even Buffon had suggested. Some thought it extended back infinitely far and may have had cyclic upheavals that led to the rising and sinking of continents, while others thought it was not infinitely old but just extremely old, so old that they were not that interested in pinning down an actual age or thought that it could even be done. They assumed that sufficient time was available for their model of a slow rate of tiny changes producing large effects to work.

In the early 1800s, geology became recognized as a formal professional discipline. Societies were formed (the Geological Society of London, the oldest national geological society, was established in 1807), university professorships were created, systematic surveys began to be done of geological strata, detailed classifications made, and studies published. It was during this period that Charles Lyell (1795–1875) published the first volume of *Principles of Geology* (1830). His book was a best seller, 15,000 copies being purchased in its first edition alone and going through ten subsequent editions, and he emerged as a leader in this new field. Significantly, as we will see later, Charles Darwin was aware of his work and they became friends later on.

Lyell too argued in favor of the uniformitarian position that the Earth was extremely old, old enough to be indeterminate even if not infinite, sufficient to produce all geological features through the process of very small but cumulative changes. He additionally argued that the present *rate* of geological change could be assumed to have been *constant over time* and thus could be used to extrapolate backward to find out when specific geologic features began to be formed. As Lyell put it, "the present is the key to the past" (Jackson 2006, 130), and that this implied a sort of steady state for a largely unchanging Earth. Lyell and other uniformitarians were successful in persuading all but the most biblically committed that the Earth was far older than earlier, purely textual, studies had estimated, and by around 1850 this idea was predominant though not unchallenged. The main debate was whether the present rate and intensity of geological events had remained the same for all time or whether those could have been greater in the past, thus shortening the estimates of the Earth's age (Rudwick 2014, 171).

By now, Enlightenment values had taken firm hold, science and rationality were on the rise, and religion could no longer rely on dogmatic assertions to suppress ideas that it found unpalatable, such as that of a very old Earth. So the strategy of those who wished to preserve the idea of an Earth that was a few thousand years old shifted to creating alternative narratives that had a scientific veneer that would make their religion-based conclusions more acceptable. The subsequent debate is illustrative for the purposes of this book because it shows how *it is always possible to construct alternative theories to salvage one's beliefs or to promote a particular agenda.*

Although this particular example involves those with a religious agenda, they are by no means unique in adopting this strategy and secular groups with an economic or political agenda have followed similar approaches. If you cannot dethrone science as the main source of reliable knowledge, the next best thing is to try to undermine trust in its conclusions by appearing to use science itself, or at least particular features of it, to advance a conclusion that is outside the scientific consensus. This is a practice that has continued down to the present day in the shape of some arguing that climate change is either not happening or is not caused by human activity, that vaccines cause autism, that smoking is harmless, and so on. The specific issues and groups may change but the pattern remains the same. It is important to not summarily dismiss these groups just because their conclusions lie so far outside the scientific consensus. Their ideas are accepted by large swathes of people and one must understand their reasoning in order to better counter them.

(d) The impact of old Earth theories on religion

In reaction to the rise in the mid-nineteenth century of uniformitarianism in geology and its concomitant idea of an extremely old Earth with possibly no beginning at all, there was a resurgence of biblical literalism that manifested itself in an alternative school of thought known as Neptunism or Flood Geology, that argued that water was the main cause of changes in the Earth's features. Because it was no longer sufficient to appeal to the authority of the Bible, this theory was advanced by those seeking to convince people that Bible-based estimates for a very young age of the Earth had a scientific justification. Some of the adherents of Neptunism were convinced that the Great Flood of Noah was sufficient to create the major geological features and thus preserve the biblical chronology; consequently this group steadfastly rejected any attempts to make the Earth older than 6,000 years or so.

One of the most well-known proponents of this theory was George McCready Price (1870–1963), who tried to make the case that *scientific* evidence supported a strictly literal interpretation of the Bible. The numbers of people who supported Price and his Flood Geology theory remained small until the rise of the creationist movement in the United States that was facilitated by the publication in

1961 of the book *The Genesis Flood* that built on Price's ideas (Whitcomb and Morris 1961). While John Whitcomb was a theologian, Henry Morris had a doctoral degree in hydraulic engineering with minors in geology and mathematics. He later founded the Institute for Creation Research in 1970 to advance these ideas, and that institution still exists with the same goals.

Even though scientists in the late eighteenth and early nineteenth centuries were often religious, most seemed to accept the idea of an extremely old Earth and were willing to concede that a strictly literal acceptance of biblical chronology was too restrictive and required the existence of unrealistically strong forces to create its effects in such a short time. To meet the need to resolve the contradiction between religious and scientific beliefs, religious apologists who were more sophisticated than the Flood Geologists now sought to find ways to reconcile an extremely old Earth with their religious texts, something that happens repeatedly when scientific evidence contradicts religious beliefs. But while Price and others sought to change science to agree with the Bible, these new attempts went in the opposite direction and consisted of inventing new interpretations of the scriptures that were consistent with scientific estimates of an old Earth. After doing so, some even went further and argued that this new agreement showed that the Bible was correct because it *predicted* that the Earth was old.

One version of this new biblical interpretation is what is known as the "Gap" (or "Ruin-Reconstruction") theory that arose in the early nineteenth century. This theory claimed to find a "gap" between Genesis 1:1 ("In the beginning God created the heavens and the earth") and Genesis 1:2 ("Now the earth was formless and empty, darkness was over the surface of the deep, and the Spirit of God was hovering over the waters") that allowed for an indefinite amount of time, and that gap was used to insert a very old, unspecified age of the universe in which matter was first created, followed by *nonhuman* life and the formation of fossils. This gap allowed for multiple cataclysms and is flexible enough to accommodate most geologic evidence. But when it comes to the first appearance of *humans*, the model reverts to that of strict standard creationism with a Garden of Eden and the first humans Adam and Eve created in six 24-hour days in 4004 B.C.E. followed in 2348 B.C.E. by Noah's flood, which in this model need not be a global flood but could be a local phenomenon, and so on (Numbers 1992).

A different reinterpretation of the Bible had an even more flexible structure and is known as the "Day-Age" model. This allows for a very old, unspecified age of the universe in which matter was first created, followed by life, the formation of fossils, and finally human beings. Noah's flood is still a historical event in this model but it could be a local phenomenon. The six "days" of creation in the Genesis story are now interpreted allegorically as representing long but indeterminate ages in time, whence comes the name of this model, and hence all the major events in Genesis have unspecified dates that can accommodate values obtained using the standard dating techniques of science. Adam and Eve and the Garden of Eden story are also interpreted metaphorically and not as actual historical people and events (Numbers 1992).

Nowadays one rarely finds people who believe in the Gap model. Christians, especially in the United States, seem to be either young-Earth Flood Geologists or some form of Day-Agers. Islamic creationists and some of the more sophisticated Christian apologists, including those in the so-called intelligent design movement, also adopt variants of the Day-Age model, because it enables them to finesse some of the contradictions between science and religion. William Jennings Bryan, a key player in the Scopes trial of 1925 that challenged the teaching of evolution in American schools, seemed to be a believer in the Day-Age model but under questioning during the trial, possibly for tactical reasons related to the specifics of that case, responded as if he was a believer in the more restrictive Gap model (Larson 1997; Singham 2009, 35–52).

(e) The impact of geology, politics, and religion on evolutionary ideas

In the early to mid-1800s, England was in a reaction against the radicalism and turmoil following the French revolution of 1789 that had dethroned the religious hierarchy there. The Tories (who later became the Conservative Party) were strong supporters of the authority of the King, the Anglican Church, and traditional biblical teachings of special creation. They were ascendant over the Whigs (who later became the Liberal Party) who wanted "extended suffrage, open competition, religious emancipation (allowing Dissenters, Jews, and Catholics to hold office) and the abolition of slavery" (Desmond and Moore 1991, 24).

The rising popularity of the idea of transmutation of species proposed by Jean-Baptiste Lamarck (1774–1829) and the spread of other radical ideas from the European mainland were undermining the idea of special creation and fueling atheistic ideas, along with generating calls for a radical restructuring of society in England. The working classes perceived the established Anglican Church in England, that lived in luxury like the Roman Catholic Church in France, as an oppressor and were calling for its abolition or at least its dethroning from its positions of privilege.

This was clearly unpalatable to those who sought to preserve a role for religion in understanding the workings of the natural world and they proceeded to try to find ways to discredit the theory of evolution, efforts that continue to this day. This was because the theory of evolution not only undermined the idea that all organisms, including humans, were specially created by God, it spawned even more dangerous ideas such as that the mind and consciousness were not separate entities but were purely creations of the brain. In a Tory-dominated climate, these raised disturbing questions such as "If life was not a supernatural gift, if the mind was not some incorporeal entity, what became of the soul? With no soul, no after-life, no punishment or reward, where was the deterrent against immorality? What would stop the downtrodden masses from rising up to redress their grievances?" (Desmond and Moore 1991, 38). Science entered into this contentious debate because the idea that species were immutable had been used to support the hereditary power of the elites. The idea that God had specially created species once for all time was used to imply that social classes were also fixed and ordained by God, and that to challenge those ideas was to challenge God's plan.

The argument that all living organisms had to have been specially created has a venerable history and its best-known formulation was provided by Christian apologist William Paley (1743–1805). Paley was an Anglican priest and his highly influential book *Natural Theology* (Paley 1802) begins with the watch-maker argument that is indelibly linked with his name and still invoked in various forms by religious apologists.

> In crossing a heath, suppose I pitched my foot against a *stone*, and were asked how the stone came to be there, I might possibly answer, that, for anything I knew to the contrary, it had lain there for ever; nor would it, perhaps, be very easy to show the absurdity of this answer. But suppose I had found a *watch* upon the ground, and it should be inquired how the watch happened to be in that place, I should hardly think of the answer which I had before given, that, for anything I knew, the watch might have always been there. Yet why should not this answer serve for the watch as well as for the stone? For this reason, and for no other, viz., that, when we come to inspect the watch, we perceive (what we could not discover in the stone) that its several parts are framed and put together for a purpose, e.g., that they are so formed and adjusted as to produce motion, and that motion so regulated as to point out the hour of the day; that if the different parts had been differently shaped from what they are, of a different size from what they are, or placed after any other manner, or in any other order than that in which they are placed, either no motion at all would have been carried on in the machine, or none which would have answered the use that is now served by it. . . . the inference we think is inevitable, that the watch must have had a maker: that there must have existed, at some time and at some place or other, an artificer or artificers who formed it for the purpose which we find it actually to answer, who comprehended its construction and designed its use. (Paley 1802, chapter 1)

For Paley, where this argument led was clear and he unambiguously stated it many pages later: "The marks of *design* are too strong to be gotten over. Design must have had a designer. That designer must have been a person. That person is GOD" (Paley 1802, chapter 23). But that confidence was to soon start to unravel because his book's publication coincided with the rise of uniformitarianism, which enabled the theory of evolution to become viable. The ideas of evolution that began to be advanced in the early nineteenth century argued that even complex organisms could come about by natural processes. In other words, the biological equivalent of a watch could emerge without requiring conscious design or a designer.

The importance and timing of the rise of uniformitarianism in geology arguing for a very old Earth cannot be overemphasized because it illustrates so clearly how *interdependent* scientific theories are and how one field of science can strongly influence other fields. On the surface, geology and biology seem to be almost independent fields of study, the former dealing with the inanimate Earth and the latter with living things. But it was during this same period of major geological advances (1830 to 1860) that Charles Darwin (1809–1882) and Alfred Russell Wallace (1823–1913) were working out their theory of evolution by natural selection, and developments in geology had major implications for their biological theories because they needed long times for it to be viable.

Darwin had at one time considered becoming a clergyman, not necessarily out of a deep sense of vocation but because it seemed to promise a respectable and comfortable life that would allow him to live in the country and give him plenty of time to indulge his true passion of studying nature, particularly beetles. During his own formal education, he had been taught Paley's ideas of special creation and had no trouble accepting them. But just prior to leaving on his voyage round the globe on the *Beagle* that lasted from 1831 to 1836, he was given the first of the three volumes of Lyell's *Principles of Geology* published in 1830 that argued for an Earth that was extremely old, so old as to make its age indeterminate. The second volume was mailed later to reach him when he arrived in South America. It is clear that Lyell's books deeply influenced him as he began speculating about alternative causes for the existence of different species. His observations of the patterns of variation of living things around the world started undermining his belief that animals had been specially created to meet the environmental niches in which they existed. Later in life, Lyell became Darwin's good friend and supporter although his own religious beliefs prevented him from fully accepting the godless implications of evolution by natural selection.

The influence of contemporary religious, political, and geological development's on Darwin's thinking is clear. He was aware of all the religious debates swirling around him as a young man, although they did not seem to divert him from his study of nature. Darwin and Wallace knew that their proposed mechanism of tiny incremental changes in organisms cumulatively leading to major changes had to be a very slow process and thus required long times to produce the organisms we see all around us, though they could not pin down exactly how much time would be needed. Although the fossil record was sparse, the fact that the pattern of fossils corresponded to their location in successive layers of geological strata linked biological evolution to geological evolution. The time taken for those strata to be formed and the intervals between the various geological epochs would parallel the time taken for evolutionary changes to occur in the fossils found in those strata. It was clear that their theory could not work with a very young Earth of 6,000 years. The rise to dominance of uniformitarian thinking in geology (which argued for a similar mechanism of small incremental changes producing big effects) and the almost unlimited time of existence for the Earth that this model projected would have been enormously liberating for both of them. It opened up a "large bank of time to draw upon" (to use a popular metaphor of that period) that allowed them to explore their ideas unfettered by time constraints that would have been inhibiting even just a few decades earlier.

The break from biblical chronology triggered by the rise of uniformitarianism also allowed for the emergence of other methods for determining the age of the various geological features. These involved measuring the rates of various processes, such as the sedimentation forming the Earth's crust and river deltas, the increase in salinity of lakes and oceans, the slowing down of the Earth's rotation due to tidal friction, and the erosion creating cliffs, ravines, and other steep formations. By calculating the rates at which these phenomena were occurring now, assuming that those rates had stayed reasonably steady over time, and making some assumptions about what their initial states might have been (such as that lakes and oceans may have originally been free of these solutes), people arrived at estimates for the time taken to reach the current state.

All of these methods of calculation gave ages in the ranges of tens of millions and even hundreds of millions of years. Although the different methods resulted in quite wide disagreement, this was not a cause of major concern because people realized that there were huge uncertainties in making estimates of the current rates at which these processes were occurring and a serious lack of precise knowledge about the initial conditions, and thus estimating the age of geological features was fraught with imprecision. So while efforts were pursued to try and resolve the discrepancies, they did not really constitute a crisis. But what was becoming resoundingly clear was that these calculations were all inconsistent with a 6,000-year-old Earth. The Bible had become irrelevant in dealing with this question.

In addition, these calculations merely gave the time taken for each of these particular features to appear and thus could be considered as only providing *lower limits* for the age of the Earth as a whole, not its actual age, so natural selection had even greater freedom to do its slow work in the creation of species, relatively unconcerned by upper limits to the time available. Although he could not pin down an actual number for the time he needed, Darwin knew that there was little evidence of major changes in species in the few thousands of years of recorded history. He also knew from his own experience with pigeon breeding that it took quite a long time to produce desired features even under strict controls and thus that an *unguided* process like natural selection would take many orders of magnitude longer, requiring at least hundreds of millions of years. But the pattern of evolution observed in the fossil record embedded in successive geological strata linked evolutionary time with geological time and he was reassured that there was plenty of evidence that the latter was "incomprehensibly vast."

> It is hardly possible for me even to recall to the reader, who may not be a practical geologist, the facts leading the mind feebly to comprehend the lapse of time. He who can read Sir Charles Lyell's grand work on the Principles of Geology, which the future historian will recognize as having produced a revolution in natural science, yet does not admit how incomprehensibly vast have been the past periods of time, may at once close this volume. . . . A man must for years examine for himself great piles of superimposed strata, and watch the sea at work grinding down old rocks and making fresh sediment, before he can hope to comprehend anything of the lapse of time, the monuments of which we see around us. (Darwin 1859, 282)

As an aside, it is not commonly known that Darwin was also an accomplished geologist, developing a keen interest in that subject while still a student and making extensive investigations while undertaking long hikes over the English countryside. Before he published his major work in biology in 1859, he had under his belt a successful theory on the origins of the coral reefs of volcanic islands, and had published three books on geology: *The structure and distribution of coral reefs* (1842), *Geological observations on the volcanic islands visited during the voyage of H.M.S. Beagle* (1844), and *Geological observations on South America* (1846), all based on his careful observations during his epic five-year voyage.

Spurred by this openness to the idea of a very old Earth, in the first edition of his *Origins* published in 1859, in one passage Darwin abandoned his usually cautious data-driven approach of argumentation and made a breezy estimate of the time taken for the "denudation" (erosion) of a region in Sussex in the south of England known as the Weald, arriving at a figure of 300 million years (Darwin 1859, 285–86). This figure must have been reassuring to him, suggesting that he did not have to worry about whether the mechanism of natural selection had sufficient time to work. But the strong criticisms that his Weald calculation received would cause him considerable grief and this particular claim got watered down in the second edition of his book and disappeared altogether in third edition published in 1861.

What this episode shows is that scientific theories are not created in a vacuum. They are strongly influenced by historical and contemporaneous factors, be they scientific, political, social, or religious. Darwin and Wallace were fortunate that they were developing their revolutionary theory at a time when the Earth was being seen as very old, the power of religious dogma to limit scientific inquiry was on the wane as science became increasingly decoupled from religion, and the intellectual climate was congenial toward what would in earlier times have been condemned as heresy and perhaps even led to death. Darwin was also fortunate to be the son, grandson, and brother of freethinkers, thus providing him with a personal space of intellectual freedom that allowed him to develop his heterodox views on the origin of species. The fact that his wife was devoutly religious would have been a hindrance to him speaking about the implications of his theory openly at home, but theirs was a warm and close relationship and she seems to have confined her concerns to worrying about his physical well-being and the fate of his immortal soul and not try to influence him in his scientific work.

(f) The backlash to the theory of evolution

The theory of natural selection advocated by Darwin and Wallace was first presented simultaneously in two papers at a meeting of the Linnean Society in 1858 and aroused little comment. A year later saw the publication of Darwin's more expansive views on the subject in the book that is now known as *The Origin of Species*, although the full title of the first edition was the wordier *On the Origin of Species by Natural Selection, or the Preservation of Favoured Races in the Struggle for Life*. While theories of evolution had been around for some decades before, his and Wallace's theory was the first to lay out a plausible mechanism for how it worked. Darwin's *Origins* was an immediate success and within a decade of its publication had gone through several editions and persuaded large segments of the scientific community that evolution was a fact and that its basic idea, that all organisms including human beings were the result of descent with modifications from common ancestors, was correct.

Interestingly, *Origins* did not explicitly include human beings in the chain of evolution. Darwin fleshed out the human connection more completely only in his later book *The Descent of Man, and Selection in Relation to Sex* published in 1871. But Darwin had hinted at this possibility at the very end of his *Origins* book where he speculated on the many future directions of research that were opened up by his theory, saying that "Light will be thrown on the origin of man and his history" (Darwin 1859, 488). This aside did not go unnoticed and was seized upon by critics who said that the idea that human beings were the mere descendants of other animal species was outrageous, and a sufficient reason for why the theory should be rejected.

Darwin's book was published at a time when it seemed that he and Wallace had all the time they needed, at least many hundreds of millions of years, for their theory of natural selection to work. But that window of freedom of long geological times was soon to begin closing and, starting around 1860, newer estimates of the age of the Earth started to shorten it considerably. Part of the reason for the decrease was undoubtedly the backlash to the success of Darwin's book in promoting the theory of evolution.

Acceptance of evolution did not mean that people had given up on a role for God altogether. Most people, including many in the scientific community, were still religious and sought to find ways to have God be part of the process. While they accepted evolution as a fact, what these people found difficult to stomach was Darwin and Wallace's *mechanism* of natural selection as the main driving force for the process, because this implied that evolution is an *unguided* process. Especially troubling was its implication that humans were, just like every other species, a mere accidental byproduct of the process and not guaranteed to appear, and thus could not be part of any original design plan that God would have had. How could the claim that humans were special and created in God's image be sustained if there was no evidence for any conscious design in the process that created them?

However, Darwin and Wallace's theory was not the only game in town. There were already competing evolutionary models around that preserved the idea of design. One alternative was *theistic evolution* in which God guided the process of evolution in unspecified ways by intervening at key moments to influence its direction and thus could guarantee the eventual emergence of humans. Another was *orthogenesis* in which it was postulated that organisms contained *within themselves* some kind of directional mechanism inserted by God that resulted in continuous improvement, leading eventually to humans as the culmination of the process. A third was *Lamarckism* in which the characteristics acquired by an organism during the course of its life were somehow transmitted to its progeny. All these alternative mechanisms were competing with natural selection for acceptance. Since they all allowed for the possibility of a role for God, or at least promised some *directionality*, they were favored by religious scientists. Natural selection offered no such assurance of a special status for humans as being the inevitable result of a grand plan.

What religious people realized was that guided processes required less time to create complex organisms than unguided ones like natural selection and so the smaller the value arrived at for the age of the Earth, the more likely it was that natural selection could not serve as the mechanism for evolution. A sufficiently young Earth, even if nowhere near 6,000 years old, would imply the need for some agency to speed up the process of evolution by guiding it along more fruitful paths. What could that agency be other than God? And so there was a strong motivation among some religious scientists to hope for a lower upper limit on the age of the Earth because that would squeeze out natural selection as a viable candidate and strengthen the case for alternative guided mechanisms.

It should be noted that natural selection is not an entirely random or chance process as some of its critics charge. While mutations do occur by chance, environmental and other pressures influence and direct the process by which some variations take hold and expand their numbers while others die out, thus the name *natural* selection to contrast with *artificial* selection where people selectively breed plants and animals to achieve specific results. But it is undoubtedly true that natural selection requires longer times than a consciously guided process to produce the end results. It is like finding one's way through a maze. One way to navigate it is at each branch point to choose a direction randomly and, if it turns out to be a dead end, backtrack to the intersection and try a different one. Using that method, which is a rough analog of natural selection, one could eventually find one's way out. But it would be much quicker if one had a map of the maze that told you which way to go, or followed some systematic plan such as walking along while always touching the wall on your left (or alternatively on your right) and thus could eliminate much wasted time backtracking or otherwise exploring dead ends.

It must be emphasized that this search for a younger Earth was a far cry from the efforts of present day creationists to resurrect a 6,000-year-old Earth. Even the religious scientists of that time had by then rejected as absurd that idea with its associated catastrophic models of geological features, and no serious scientist for the past two hundred years has considered that possibility seriously (Burchfield 1975, 37). Anyone who now argues for such a young Earth can be said to have rejected science altogether. These earlier religious scientists accepted evolution and an old Earth as facts. What they were hoping for is that their methods of estimating the age of the Earth using purely scientific methods would yield results that were of the order of less than a hundred million years, which would be too tight a timescale for natural selection to work.

It was Lord Kelvin who seriously threatened to overthrow the natural selection mechanism for evolution. Kelvin had already established a reputation as one of the foremost physicists of his day, pioneering important work in many areas but especially in the field of thermodynamics, particularly with the first law that dealt with the conservation of energy, and the second law that dealt with the dissipation of heat energy and the consequent directionality of heat flow. His words were taken seriously, since in those times physics, along with astronomy and mathematics, were seen as the scientific disciplines on the firmest footing.

Kelvin was opposed to the idea of natural selection because it seemed to rule out any role for a designer of the universe, but that was not his only reason for challenging Darwin. He also felt that the long ages demanded for natural selection to work were unrealistic (Burchfield 1975, 33, 72). He was also strongly opposed to uniformitarianism because he felt that the idea of processes remaining unchanged over long, indefinite periods of time went against the second law of thermodynamics (which he had been instrumental in elucidating) that things change and dissipate over time. What Kelvin did was to apply the laws of physics to three different kinds of calculations for the age of the Earth: calculating the age of the Sun to set an upper limit to the solar system; examining the role that tidal friction played in shaping the Earth; and treating the Earth as a solid cooling body. All three methods gave him similar results, persuading him that he was on the right track (Jackson 2006, 200).

In the last method involving cooling, Kelvin assumed the same basic model as Buffon had in 1778, specifically that all the Earth's (and Sun's) energy originated as gravitational and mechanical energy in the particles and possibly meteors that preceded its formation, and that as they coalesced to form the Earth, this energy was transformed into heat energy that made the Earth into a hot molten ball. The Earth, following the laws of thermodynamics, gradually cooled and solidified as this heat was slowly radiated away into space. What had changed since Buffon's time was that science's understanding of the laws of physics, especially thermal physics, had advanced considerably, and Kelvin himself had played a major role in that advancement. Also, technological improvements in thermometers enabled far more accurate measurements of temperature than Buffon could make using just the sense of touch.

Knowing the mass of the Earth and the Sun and making various assumptions about the initial kinetic energy of the particles, the state of the Earth's core, and the rates of conductivity of heat from that core to the surface and subsequent radiation into space, in 1862 Kelvin came up with upper limits for the ages of both the Earth and the Sun, captured in his statement that "It seems, therefore, on the whole most probable that the sun has not illuminated the earth for 100,000,000 years, and almost certain that he has not done so for 500,000,000 years." He also estimated that the Earth was somewhere between 20 million and 400 million years old, with the likely figure being 98 million (Burchfield 1975, 36). In 1868, he revised his calculations and made an even stronger statement, that the Earth was no more than 100 million years old (Burchfield 1975, 43). He thus flatly contradicted Darwin's calculations of 300 million years for the time taken for the denudation of the Weald, though by that time that calculation had disappeared from the third and subsequent editions of Darwin's book due to other criticisms he had received.

Kelvin had thrown down the gauntlet to the geologists and biologists who were working with much longer times and the next fifty years would result in a struggle among physicists, geologists, and biologists to pin down the age of the Earth and see which side would emerge as the winner: the young Earth scientists of that time (which then only meant ages of less than 100 million years, not 6,000 years) or the old Earthers. It should be emphasized that while Kelvin disliked the theory of evolution by natural selection because it seemed to rule out design in nature and thus left no role for God, and even though his research seemed to support the idea that there was insufficient time for it to be the mechanism for evolution and was thus congenial to him, there is no evidence that he deliberately distorted his work to get the results he preferred (Burchfield 1975, 33).

(g) The three-way conflict of physics, biology, and geology

While Kelvin's estimates of the age of the Earth were interesting in their own right and faced their own supporters and detractors, the undeniably important consequence of his work was that for the first time, pinning down a fairly precise age for the Earth became a question that had the potential to be definitively answered, using principles and laws that had been tested and established independently, and this spurred the growth of an entire research area. Kelvin did two things that were of immense importance. The first was that he established the importance of doing precise calculations instead of the uniformitarians' vague notions of a large and possibly limitless time for the age of the Earth. The second was that he highlighted the importance of *unifying scientific theories* from seemingly independent fields, by introducing physics principles into geological studies. This illustrates an extremely important point that I will return to later, that the theories of science form an interconnected web in which the validity of each one cannot be tested independently but only within the context provided by other theories. In research into the age of the Earth, we see major theories of biology, geology, chemistry, paleontology, and physics being explicitly and inextricably linked.

By using such modern methods, Kelvin had arrived at a value for the age for the Earth that was so low that it made it very difficult for natural selection to be its prime mechanism. Then, as now, results from physics tend to be regarded as on a firmer footing than those from other disciplines because they can be obtained under more controlled experimental considerations and the theories used seem to be more tightly constrained by data. Hence other fields such as geology and paleontology and biology tend to try to conform to the constraints provided by physics, not the other way around. Kelvin's estimate of 100 million years as the upper limit for the age of the Earth became part of the scientific lore and geologists and biologists scrambled to accommodate it by trying to find ways to modify their own theories to be consistent with this upper bound.

Geologists had some success since their calculations depended on many parameters whose values could not be determined precisely and so there was considerable room for flexibility. But for biologists advocating for evolution by natural selection, this low upper limit for the age of the Earth caused serious problems. Wallace had published a book in 1880 that suggested that 200 million years was sufficient for evolution to have worked (Jackson 2006, 193). By squeezing here and pinching there, it seemed possible to accommodate 100 million years as just barely sufficient for natural selection to work, but only with great difficulty and at the risk of sacrificing plausibility.

By around 1880, an uneasy truce seemed to have been drawn amongst the physics, geology, and biology research communities around a value of 100 million years for the age of the Earth. But that tentative consensus did not last long. Other scientists came along who followed up on Kelvin's methods using more refined calculations and newer estimates for the parameters involved, and they arrived at even shorter ages, first 40 million and then 20 million years. Most important amongst these was an 1893 calculation by Clarence King, who had served from 1879 to 1881 as the first director of the US Geological Survey who, again basically using Kelvin's thermal methods, arrived at a figure of 24 million years.[1]

Kelvin himself, in a paper in 1897 toward the end of his long and illustrious career, stated his conclusion that the Earth was between 20 and 40 million years old, with King's value of 24 million being the most likely to be correct. What made Kelvin's conclusions about his new lower values carry great weight was that he had also done calculations for the age of the Sun and found that the gravitational collapse of the Sun would only be able to provide enough energy to warm the Earth for about 20 million years. Since the sedimentation process used by geologists as one of their dating methods for the Earth depended on the warmth of the Sun, this set a more stringent limit on the age of the Earth than even Kelvin's terrestrial calculations (Stacey 2000).

---

[1] I feel compelled to bring to the reader's attention a fascinating and incredible personal story about King even though it has nothing to do with anything discussed in this book. In *Passing Strange: A Gilded Tale of Love and Deception Across the Color Line*, historian Martha A. Sandweiss tells us how this blue-eyed, fair-skinned, white man who "dined at the White House, belonged to Manhattan's most elite clubs, and parlayed his privileged upbringing and Ivy League education into a career as an eminent scientist, writer, and government official" (Sandweiss 2009, 8) also, for the last thirteen years of his life, lived a secret double life as a black Pullman porter named James Todd. He took on this alternate identity in order to marry the woman he loved, Ada Copeland, a former slave, and fathered five children with her. He somehow kept the secret of his double life from everyone, including her, revealing all only in a letter written to her on his deathbed in Phoenix in 1901.

These new results caused immense problems for the other areas of science. If Kelvin's results were correct, then almost all of geology would have to be drastically reconceptualized. For example, the geologist John Joly had estimated that it had been about 90 to 100 million years since the Earth's crust had cooled sufficiently to allow water to condense and the oceans to form (Rudwick 2014, 233). But this time geologists were not as willing to concede ground to the physicists. They seemed to have had enough of conforming to the increasingly restrictive limits placed on the age of the Earth by the physicists and modifying their parameters accordingly. Their discipline had now been around for about a hundred years and this later generation of geologists could not be intimidated as easily by the older physics discipline. They felt that 100 million years was as far as it was reasonable for them to go given their own methods of estimating the ages of geological features based on rates of formation and erosion and sedimentation. They dug in their heels and became more assertive, saying that the laws of geology were firmly enough established to rule out such a young Earth, and boldly suggested that it was physics that had gone awry somewhere, even if they could not find fault in its calculations or point out where the problematic assumptions were.

In the case of biology, resistance to the developments in physics that argued for a 20 million year Earth was much weaker. If Kelvin were right, the theory of evolution by natural selection would have to be thrown out the window, to be replaced by some teleological model of guided evolution that implied planning or design or some other form of supernatural intervention to speed the process along.

Darwin and Wallace's theory of natural selection had already been in retreat during the latter half of the nineteenth century under the assaults of both physicists and some biologists. Fleeming Jenkin (a physicist, engineer, and collaborator of Kelvin) had, in an anonymous review of *Origins*, delivered a severe critique of Darwin's theory that focused on two major points (Jenkin 1867). The first was that the then-dominant theory of "blending inheritance," which said that children had a mixture of the qualities of their parents, worked against Darwin's theory, since even if one parent experienced an advantageous mutation, that person's child would only have half of it because the other parent would not have it, the grandchild would have just one-fourth, and so on. The mutation would thus get diluted and disappear over time and not grow and dominate the population as natural selection argued.

Jenkin fleshed out his argument against blending inheritance with an example that I will quote at length because it is also reflective of the beliefs of his time that saw white people as unquestionably superior to others. (Jenkin uses the word "sport" to describe a rare mutation that results in a dramatic variation.)

Suppose a white man to have been wrecked on an island inhabited by negroes, and to have established himself in friendly relations with a powerful tribe, whose customs he has learnt. Suppose him to possess the physical strength, energy, and ability of a dominant white race, and let the food and climate of the island suit his constitution; grant him every advantage which we can conceive a white to possess over the native; concede that in the struggle for existence his chance of a long life will be much superior to that of the native chiefs; yet from all these admissions, there does not follow the conclusion that, after a limited or unlimited number of generations, the inhabitants of the island will be white. Our shipwrecked hero would probably become king; he would kill a great many blacks in the struggle for existence; he would have a great many wives and children, while many of his subjects would live and die as bachelors; an insurance company would accept his life at perhaps one-tenth of the premium which they would exact from the most favoured of the negroes. Our white's qualities would certainly tend very much to preserve him to a good old age, and yet he would not suffice in any number of generations to turn his subjects' descendants white. It may be said that the white colour is not the cause of the superiority. True, but it may be used simply to bring before the senses the way in which qualities belonging to one individual in a large number must be gradually obliterated. In the first generation there will be some dozens of intelligent young mulattoes, much superior in average intelligence to the negroes. We might expect the throne for some generations to be occupied by a more or less yellow king; but can any one believe that the whole island will gradually acquire a white, or even a yellow population, or that the islanders would acquire the energy, courage, ingenuity, patience, self-control, endurance, in virtue of which qualities our hero killed so many of their ancestors, and begot so many children; those qualities, in fact, which the struggle for existence would select, if it could select anything?

Here is a case in which a variety was introduced, with far greater advantages than any sport ever heard of, advantages tending to its preservation, and yet powerless to perpetuate the new variety. (Jenkin 1867, 289–90)

The theory of inheritance based on what we now call genetics proposed by the scientist and monk Gregor Johann Mendel (1822–1884) would have countered Jenkin's objection and saved Darwin's theory. But Mendel's work had been published in 1866 in a relatively obscure journal and was thus not widely known. Mendel had obtained and read Darwin's books as soon as they were translated into German. He had sent reprints of his paper to many scientists in the field, including Darwin, Wallace, and Jenkin, but they all seemed to be unaware of it, even though Mendel's reprint was later found among Darwin's papers (Henig 2000, 143).

Darwin had in fact already anticipated the difficulty posed by blending inheritance and pre-emptively suggested that if the mutation occurred independently in several people and they mated with each other, then the dilution of the advantage would not occur and the population could grow. But he knew that this was a weak argument. In fact, Darwin had never even argued that natural selection was the *only* mechanism at work in evolution. Even in the first 1859 edition of *Origins*, he allowed for some Lamarckian ideas, such as environmental influences shaping evolution, and the use–disuse mechanism, whereby a feature that was used heavily and was thus highly developed in an organism would be passed on in an enhanced form to the progeny, while one that was used less would be weaker in the progeny. He suggested that this could explain the sightless eyes of organisms that occupy deep underground caverns where no light penetrates and thus eyes were of no help, and the flightless birds found on islands where they had no predators and thus no reason to fly (Darwin 1859, 454). But as a result of the challenges to natural selection, by 1880 he was willing to concede a greater role for such alternative mechanisms than he had done in the past.

The second argument of Jenkin seemed to trouble Darwin more and that was the estimate that Jenkin made about the time needed for natural selection to work. Jenkin was a friend and business partner to the older and more distinguished Kelvin, and some historians have characterized the relationship as Jenkin's Dr. Watson to Kelvin's Sherlock Holmes (Morris 1994). Jenkin used Kelvin's arguments to point out that even carefully controlled and selected breeding produces only minute changes over long periods and that natural selection's need for indefinitely long times in the past were unjustified because "past ages are far from countless." He used Darwin's own words against him.

Darwin says with candour that he 'who does not admit how incomprehensibly vast have been the past periods of time,' may at once close his volume, admitting thereby that an indefinite, if not infinite time is required by his theory. Few will on this point be inclined to differ from the ingenious author. We are fairly certain that a thousand years has made no very great change in plants or animals living in a state of nature. The mind cannot conceive a multiplier vast enough to convert this trifling change by accumulation into differences commensurate with those between a butterfly and an elephant, or even between a horse and a hippopotamus. A believer in Darwin can only say to himself, Some little change does take place every thousand years; these changes accumulate, and if there be no limit to the continuance of the process, I must admit that in course of time any conceivable differences may be produced. He cannot think that a thousandfold the difference produced in a thousand years would suffice, according to our present observation, to breed even a dog from a cat. He may perhaps think that by careful selection, continued for this million years, man might do quite as much as this; but he will readily admit that natural selection does take a much longer time, and that a million years must by the true believer be looked upon as a minute. Geology lends her aid to convince him that countless ages have elapsed, each bearing countless generations of beings, and each differing in its physical conditions very little from the age we are personally acquainted with. This view of past time is, we believe, wholly erroneous. So far as this world is concerned, past ages are far from countless; the ages to come are numbered; no one age has resembled its predecessor, nor will any future time repeat the past. The estimates of geologists must yield before more accurate methods of computation, and these show that our world cannot have been habitable for more than an infinitely insufficient period for the execution of the Darwinian transmutation. (quotation in original, Jenkin 1867, 294–95)

In his private correspondence with friends and colleagues, Darwin expressed concerns about the problems raised by Jenkin and others about the shortness of time (Morris 1994). But Darwin was still hopeful that natural selection would eventually weather the storm caused by the short time scales, and that new future understandings about the physics of the Earth and the rate of species change might vindicate his theory. In the sixth and final edition of *Origins*, he gave his final published statement on this topic, saying:

With respect to the lapse of time not having been sufficient since our planet was consolidated for the assumed amount of organic change, and this objection, as urged by [Lord Kelvin], is probably one of the gravest as yet advanced, I can only say, firstly that we do not know at what rate species change as measured in years, and secondly that many philosophers are not yet willing to admit that we know enough of the constitution of the universe and of the interior of our globe to speculate with safety on its past duration. (Darwin 1872, 391–92)

When Darwin died in 1882, he was widely mourned as a great scientist who had convinced the world that evolution is a fact from which there is no going back and that we had a basic understanding of the mechanisms driving it. But he died with a cloud hanging over his favored mechanism of natural selection and perhaps wondering if it would survive. Vindication of his theory had to wait more than three decades while the various contradictions were wrangled over.

As mentioned earlier, by the end of the nineteenth century, geologists had started balking at the ever-reducing ages of the Earth coming out of physicists' calculations, most of the latter based on thermal cooling models going back to Kant and Laplace and Buffon and refined by Kelvin. They felt that 100 million years was as low as they were willing to go. Some physicists were also starting to question Kelvin's models and the values of the thermodynamic parameters upon which his calculations depended. Biologists were also getting tired of being pushed around by the physicists. Perhaps emboldened by the resistance from geologists and some physicists, they too started arguing that the evidence from paleontology for an older Earth than was allowed by physics was too strong to be dismissed, and that the cause of any discrepancy lay with physics and not with these other fields of research.

In a series of three papers published in 1895, one of Kelvin's former assistants John Perry showed that allowing for inhomogeneity and including convective heat flow in the Earth's interior, something that Kelvin had not allowed for when he treated the Earth as a rigid homogeneous body, could invalidate Kelvin's conclusions since they could increase the age of the Earth by factors of around 50 or even 100 over Kelvin's estimates (Perry 1895a, 1895b, 1895c). In a talk to the British Association in 1896, Edward Poulton (1856–1963), professor of zoology at Oxford University, picked up on these critiques of Kelvin's work by Perry and other physicists and made the case that the idea of an older Earth based on biological and paleontological evidence should not be so easily dismissed, suggesting that Kelvin could well be wrong about his young Earth (Burchfield 1975, 139–40).

This was the state of affairs near the end of the nineteenth century. It was an impasse, with the major theories of geology and physics at loggerheads about the age of the Earth, and the fate of the theory of evolution by natural selection hanging in the balance, to be decided by which side would emerge the ultimate winner.

(h) The revolutions in physics in the twentieth century

The dawn of the twentieth century saw an extraordinary period of ferment in the world of science. In the case of physics, in addition to the turmoil over the age of the Earth, there were other crises. The discovery of the existence of atomic spectral lines (that each atom emitted light that consisted of a discrete set of wavelengths) seemed inexplicable using the well-established theory of electrodynamics. Another crisis was that the "luminiferous aether," the supposed carrier of light waves that was believed to permeate all space, seemed to be extraordinarily successful in evading every attempt at unambiguously detecting its presence. The idea of a physical entity occupying all of space and yet being undetectable was an unpalatable idea, reeking of mysticism, and yet the aether seemed to be necessary to enable the propagation of light. In addition, the blackbody radiation spectrum seemed to defy understanding on the basis of what were thought to be well-established laws of mechanics and radiation.

But as is often the case, crises also stimulate new discoveries and inventions and that same period also produced one major scientific revolution after another. The year 1900 saw the rediscovery of Mendel's theory of genetics that showed that the heritable qualities we had were stored in our bodies in *discrete* units that were passed on *intact* to our progeny and not blended away on breeding as previously thought, thus removing one major objection to natural selection that had been highlighted by Fleeming Jenkin. The integration of Mendelian and population genetics with natural selection into what is now known as the modern *neo-Darwinian evolutionary synthesis* was, however, not fully realized until around 1930, about seventy years after Darwin's book first appeared. This is one of the many examples of how long scientific revolutions take to become firmly established.

The year 1900 also saw the introduction by Max Planck (1858–1947) of the quantum hypothesis that was able to reproduce the pattern of radiation emitted by a black body (a long-standing puzzle of physics), although in an ad hoc manner. In 1905 Albert Einstein (1879–1955) emerged with his theory of special relativity that made the aether redundant and dispensable, thus solving the mystery of why it had seemed so elusive. The aether ceased to be necessary as an explanatory concept and was duly consigned to a state of nonexistence. That year also saw the publication of a second paper by Einstein, which contained the result that is now famously rewritten as $E=mc^2$, showing that small amounts of mass could be converted to large amounts of energy. Niels Bohr (1885–1962), in 1913, introduced the planetary model of the atom consisting of negative charges orbiting like planets around a tiny but much more massive positively charged nucleus that played the role of the Sun. That model explained the discrete atomic spectral lines and was the first step in resolving the contradictions between the new atomic theory and electrodynamics.

But as far as the age of the Earth was concerned, the major discoveries that impacted it were the discovery of X-rays in 1895, followed rapidly by the discovery of alpha, beta, and gamma radiation of radioactivity. There began a rush to isolate the elements that produced this extraordinary new radiation and to measure their properties. Marie Curie (1867–1934) and Pierre Curie (1859–1906) were among the leaders of the quest in isolating these elements and soon the list contained actinium, uranium, polonium, thorium, and radium. In 1903, Pierre Curie showed that the decay of radium resulted in the emission of prodigiously large amounts of heat. An early calculation that same year showed that a mere 3.6 grams of radium per cubic meter would be sufficient to be the source of energy radiated by the Sun (Burchfield 1975, 166). Since radium was known to exist within the Earth, people immediately realized that if the heat generated by radioactive decay did not have its origins in the familiar and well studied forms of gravitational, mechanical, and chemical energy (and it soon became clear that it did not) but came from this new form, then this meant that there was a source of heat generation that had not been previously taken into account that would result in the Earth's age becoming larger.

The commonly told version of events is that the old methods of calculation of Kelvin and other physicists were undermined because of the discovery of radioactivity. But that is another one of those myth stories that physicists pass on to their students that is not borne out by the historical record, since it was not shown that enough heat would be produced by this new mechanism to significantly change Kelvin's conclusions.

> We now know that the crust does not contain enough radioactive heat to explain the surface heat flux; nevertheless, it is still frequently stated that, because the discovery of radioactive heat undermined an assumption behind Kelvin's calculation, it also undermined his conclusion. This statement is logically incorrect; Kelvin's conclusion would be undermined by that discovery only if incorporation of the Earth's radioactive heat into his calculation produced a substantially different age for the Earth. (England, Molnar, and Richter 2007)

These authors argue that it was Perry's earlier work that was decisive in removing Kelvin's restrictive lower limits on the age of the Earth. While radioactivity played a smaller role than popularly thought in undermining Kelvin's low upper limits for the age of the Earth, its great significance was, as will be discussed in the next section, that it gave birth to a new and far more accurate method for determining that age. The other serious constraint, that the Sun's warmth was limited to just 20 million years, remained in effect, but that problem was relegated to the background in the hope that it too would go away in the future, as it indeed did with the later discovery of thermonuclear fusion as the energy source for the Sun and stars.

Since there was no longer any method of precisely determining the absolute age of the Earth, one of the major barriers to the desire of geologists and paleontologists and natural selectionists for longer ages had fallen, and this development was greeted with considerable relief within those communities of scientists. It is unfortunate that Darwin did not live long enough to see the dismantling of the main impediment to his theory.

(i) Consensus emerges on the age of the Earth

It was Ernest Rutherford (1871–1937) who, along with others, is associated with some of the key conceptual breakthroughs involving radioactivity, such as realizing that during radioactive decay one unstable element was transformed into another, which in turn was transformed into another, and so on until a stable element was reached, at which point the process stopped. It was found that if we took a sample of radioactive material, the time taken for the initial amount of material to decay to half its original value was a fixed amount (known as its "half-life") that is *independent* of the amount of material we started with.

So if we start with (say) 160 grams of some radioactive element that has a half-life of (say) 10 days, then after 10 days we would have 80 grams of the original element left with the rest having decayed away into other elements. After 20 days (i.e., another 10 days) we would have 40 grams left, after 30 days, we would have 20 grams left, and so on, with corresponding increases in the amount of the products created by the decay chain. What was to prove highly significant and useful was that each radioactive element had its own signature value of half-life and there was a huge range of half-lives for the different radioactive elements, ranging from fractions of seconds to hundreds of millions, even billions, of years. This discovery proved to be the key to arriving at the current consensus on the age of the Earth.

The discovery of radioactivity and the associated concept of half-lives of elements opened up the possibility of determining the absolute age of rocks. This argument was developed by Bertram Boltwood in the very early 1900s and is the process now referred to as radiometry (Jackson 2006, 237). The argument goes like this: Suppose a sample of rock is found to contain 50 grams of a parent radioactive material P and 350 grams of the stable final daughter element D in its radioactive series, and also suppose that the half-life of this decay has been measured to be 10 million years. If we assume that the rock initially had only the parent element P and no D, then we can figure out that the rock sample must have initially consisted of 400 grams of P. Then after 10 million years, it had 200 grams of P and 200 grams of D; then after another 10 million years, it would have 100 grams of P and 300 grams of D; and after another 10 million years, it would have 50 grams of P and 350 grams of D. Hence the rock must have formed 30 million years ago.

Of course, nothing is that simple in real life. Apart from all the difficulties in finding and measuring the properties of suitable rocks, an obvious complicating factor is that some of the parent and daughter elements may have escaped or entered the sample during the period of decay, which is a real possibility during the time that the rock was molten before it solidified. The effects of this can be corrected for by the method known as *isochron dating* (Stassen 1998). If there are many different rocks with different radioactive series with multiple elements involved, and their results converge around a single age, that lends further confidence to the result. Once the absolute ages of rocks could be determined with considerable confidence, scientists could now determine, in addition to their relative ages, the *absolute* ages of the various geological layers in the Earth's crust that were formed by sedimentation. They did this by measuring the ages of the igneous rocks that were formed by volcanic activity that created molten rocks that cooled and are now buried in those layers. This provides an additional measure of the ages of the fossils that are found in those layers.

Apart from all this valuable information about the absolute ages of geological layers and fossils that were now possible to be determined, the race was also on to use these new methods to find the *oldest* rocks on the planet because those would set a lower bound on the age of the Earth, and records kept tumbling at a rapid pace. Rutherford in 1905 measured rocks that were 141 million years old and shark's teeth that were 77 million years old. In 1907, Boltwood published the ages of mineral specimens that ranged from 400 million to 2.2 *billion* years (Jackson 2006, 237). Such old ages would have been unthinkable just ten years earlier. The fact that these figures were now not summarily rejected shows how rapidly things had changed.

Geologist Arthur Holmes (1890–1965) produced a time scale in 1911 based on the rocks found in various geological strata and produced dates for the various eras such as the Carboniferous (340 million years), Devonian (370 million years), and Silurian (430 million years), values that hold up remarkably well when compared with current values of 360, 416, and 444 million years respectively (Jackson 2006, 245). The oldest rocks produced by Holmes at that time were 1.64 *billion* years old from Sri Lanka (then called Ceylon).

So in a little over a decade after the discovery of radioactivity, and just four years after Kelvin died in 1907 possibly still unconvinced that the Earth was more than tens of millions of years old (there are ambiguous reports about whether he accepted the new implications of radioactivity shortly before his death), there was powerful and persuasive evidence that the Earth was really and truly old, in the range of billions of years. It was a momentous change in thinking occurring over a remarkably short time, showing how the accumulation of evidence can result at some point in the *scientific community* changing its views on some question extremely rapidly by going where the preponderance of evidence leads, even while some individual scientists resist the changes.

These much larger times were not uncritically accepted, however. Critics charged that we could not know that the half-lives of radioactive elements had remained the same through all time. Maybe they used to decay faster in early days, misleading us as to their ages. But as more research was done, these objections began to disappear, and soon other methods (such as the presence of what are known as pleochroic halos in some materials) led to estimates of ages of rocks that were also in the hundreds of millions of years.

Oddly, some geologists who had become accustomed to modifying their theories to accommodate a 100 million-year-old Earth and had resisted efforts to lower it to 20 million, were disconcerted by the much *longer* times that were now being suggested. In a paper written following a British Association symposium

on this topic, geologist William Sollas described the dramatic nature of the transformation that had occurred and colorfully urged his fellow geologists to not get
carried away by these new developments.

> The geologist who had before been bankrupt in time now found himself sud-
> denly transformed into a capitalist with more millions in the bank than he
> knew how to dispose of.
>
> The consequences have been far-reaching; already some geologists, thus
> newly enriched . . . have begun to rebuild their science on a new and magnif-
> icent scale, while more cautious people, like myself, too cautious, perhaps, are
> anxious first of all to make sure that the new clock is not as much too fast as
> Lord Kelvin's was too slow. (Sollas 1921)

But while these signs of long ages for the Earth solved the immediate problems of
biologists, geologists, and paleontologists, and were greeted by them with great
relief, it did create a serious contradiction with calculations for the age of the
Sun, which still remained at about 20 million years. The idea of the Earth being
older than the Sun was manifestly paradoxical but it was assumed by scientists
in other fields that this was a problem for physicists to address. Radioactivity
and better models for the Earth had eliminated the immediate and acute crises
in their fields and those scientists could ignore what seemed like a more distant
and less well-understood problem. And indeed, with the detailed theoretical
work of Hans Bethe and others around 1940 (Bethe and Critchfield 1938; Bethe
1939) showing that thermonuclear fusion processes, and not gravitational collapse, could explain the composition and energy powering the Sun and other
stars, that problem too went away as the old calculations for the age of the Sun
were seen to be no longer valid (Stacey 2000). The age of the Sun is now estimated
to be about 4.6 billion years.

The minimum age of the Earth kept getting pushed to larger values as older
rocks kept being found and methods of analysis improved with the invention of
new instruments such as the mass spectrometer. The minimum age was raised
to 1.90 billion years in 1935, 3.35 billion years in 1947, and 3.45 billion years
in 1956.

But then there arose a problem from an entirely different and unexpected
direction, the new field of cosmology. In 1927 Georges Lemaître (1894–1996)
predicted, based on observations of the speeds and distances of galaxies, that the
universe was expanding and gave an equation for the rate of expansion (Haynes
2018). This was corroborated in 1929 by the observations of Edwin Hubble
(1889–1953). (The International Astronomical Union proposed in 2018 that the
law should be called the Hubble-Lemaître law rather than simply the Hubble law
as is currently the custom.) This law enabled scientists to estimate the time when

the universe would have begun, and they initially arrived at an age of 1.80 billion years (Jackson 2006, 251). It was absurd to suppose that the universe was younger than the Earth and this caused some consternation. But as the reach of telescopes increased and greater and greater expanses of the vastness of the universe came under observation, the improved quality, range, and amount of data available to calculate the age of the universe resulted in its value steadily increasing, to 10 billion years by the early 1950s, to 13 billion years by 1958, and finally to the present value of 13.8 billion years. Thus the paradox of the universe being younger than the Earth was also resolved.

While the ages of rocks provided a *lower limit* to the age of the Earth, to obtain the actual age of the Earth we would need to find rocks that were formed at the time of the Earth's formation. But the problem with finding those rocks is that plate tectonics would likely have crushed most of them or sent them deep underground so they are not easy to obtain. Despite this, determined efforts have steadily unearthed older and older rocks from all over the globe (Dalrymple 1991).

The last major shift in the age of the Earth came when Clair Patterson (1922–1995) studied the age of meteorites that had crashed to the Earth. It was assumed that these meteorites were formed at the same time as the Earth and the solar system, but since they arrived here much later, were less likely to be contaminated by the ebbs and flows of Earth's geological history. Patterson studied two samples from the Canyon Diablo meteorite that fell in Arizona about 50,000 years ago and reported in 1953 that their ages were 4.510 and 4.570 billion years (Jackson 2006, 257). More and more meteorites from all over the globe started to be analyzed and their ages converge to around 4.5 billion years. The oldest moon rocks have been measured to be between 4.4 and 4.5 billion years old, which is consistent with the other dates.

The age of the Earth is now generally accepted to be 4.54 billion years with an uncertainty of about 1%. Despite the major swings in estimates that preceded arrival at this point due to the different theories that were used, this is about as firmly held a "fact" in science as one can imagine. It has been arrived at by a convergence of theories and data and analyses from a wide range of scientific disciplines that, by the mid-twentieth century, has resulted in a model for the Earth that allows for occasional major but *local* events such as earthquakes, volcanoes, and floods to be superimposed on an overall slow and steady process (Rudwick 2014, 268). It is this *interlocking* of theories that makes it hard to envisage the value for the age changing considerably in the future, unless a major new discovery, comparable to radioactivity in its widespread impact, results in a radical re-evaluation of theories in *all* those areas.

Physicist John Strutt (better known as Lord Rayleigh) expressed the need for this kind of unified approach that seeks consistency over a wide range of disciplines.

> There has been a tendency on all sides for specialists in one branch of science to consider themselves free to disregard evidence drawn from a class of considerations with which they are not familiar. I am sure that this is not the road to truth. In attempting a problem of this kind, when we seek to plumb into the depths of time, far beyond human experience, we cannot afford to neglect evidence drawn from any quarter, even if it is not the kind of evidence which we find it most congenial to contemplate. (Rayleigh 1921)

The search for an answer to the age of the Earth is truly a remarkable story and its key features are repeated in the study of other phenomena as well. There are many lessons that the search can teach us and, as we will see in the ensuing chapters, both critics and supporters of science can find things in it to provide support for their respective causes.

7

What we learn about science from the study of the age of the Earth

[The story of how we arrived at the current age of the Earth shows that reversals of seemingly firm conclusions are the *norm* in science, not the exceptions, and form an integral part of its process. They are thus not a cause for alarm and the community of scientists has over time developed ways to arrive at consensus judgments that, while not infallible, can command considerable confidence.]

Critics of science can and do seize upon the fact that science has given different answers at different times to even straightforward questions like the age of the Earth to argue that the conclusions of science cannot be relied upon and that what is asserted confidently today could be different tomorrow. Why should we believe that the current value of 4.54 billion years is the final answer? What makes us confident that we have finally got it right after so many major swings in the value? Is it not possible that the answer might yet shift to a completely different value, perhaps even as low as 6,000 years, thus vindicating a literal reading of the chronologies in the Bible? In the United States, groups that believe this are not insignificant in number and have even been able to raise hundreds of millions of dollars to construct museums and even a replica of Noah's Ark that purportedly show the 6,000 year history of the Earth.

Another criticism that is leveled at science by critics is that scientists have their own beliefs and prejudices and are thus not free of bias, which can influence their work and their conclusions. In calculating the age of the Earth, there was considerable flexibility in deciding on what models and parameters were reasonable to use and those scientists like Kelvin who would have liked to squeeze out natural selection may have tended toward selecting those that gave them a younger Earth, while those like Darwin who wished to preserve natural selection as a viable mechanism may have tilted it in the opposite direction, even if neither side did so deliberately. We should never forget that scientists *are* human. There is always the problem of subjective biases where the scientist sees what she wants to see and overlooks or disregards things that challenge her beliefs. A well-known problem is that of *confirmation bias*, the tendency of people to seize upon, remember, and examine less critically those ideas and pieces of evidence that support their preconceived beliefs and to judge much more harshly those that go against them. Even scientists who pride themselves on their heightened levels of objectivity are not immune from this failing.

The idea of biases prejudicing and clouding one's objectivity have been around for a long time, even before the term confirmation bias entered the psychology lexicon around 1960. The Greek philosopher Thucydides (c. 460 B.C.–c. 395 B.C.) stated in his book *The History of the Peloponnesian War* that "their judgment was based more upon blind wishing than upon any sound prevision; for it is a habit of mankind to entrust to careless hope what they long for, and to use sovereign reason to thrust aside what they do not fancy" (Thucydides 1910, Book IV, Chapter XIV). Francis Bacon wrote that "the human understanding when it has once adopted an opinion (either as being the received opinion or as being agreeable to itself) draws all things else to support and agree with it. And though there be a greater number and weight of instances to be found on the other side, yet these it either neglects and despises, or else by some distinction sets aside and rejects, in order that by this great and pernicious predetermination the authority of its former conclusions may remain inviolate" (Bacon 1620, 36).

Scientists are susceptible to this failing too as Leo Tolstoy recognized when he wrote that "I know that most men—not only those considered clever, but even those who are very clever, and capable of understanding most difficult scientific, mathematical, or philosophic problems—can very seldom discern even the simplest and most obvious truth if it be such as to oblige them to admit the falsity of conclusions they have formed, perhaps with much difficulty—conclusions of which they are proud, which they have taught to others, and on which they have built their lives" (Tolstoy 1899, Chapter XIV, 124). And Herbert Spencer wrote, "Those who cavalierly reject the Theory of Evolution as not being adequately supported by facts, seem to forget that their own theory is supported by no facts at all. Like the majority of men who are born to a given belief, they demand the most rigorous proof of any adverse belief, but assume their own needs none" (Spencer 1852).

The philosopher of science Pierre Duhem was aware of this trap that scientists can fall into and said in his book *The Aim and Structure of Physical Theory* that scientists should strive to act like judges, look carefully at *all* the evidence, and try to leave their personal preferences at the door.

> Now nothing contributes more to entangle good sense and to disturb its insight than passions and interests. Therefore, nothing will delay the decision which should determine a fortunate reform in a physical theory more than the vanity which makes a physicist too indulgent towards his own system and too severe towards the system of another. We are thus led to the conclusion so clearly expressed by Claude Bernard: The sound experimental criticism of a hypothesis is subordinated to certain moral conditions; in order to estimate

correctly the agreement of a physical theory with the facts, it is not enough to be a good mathematician and skillful experimenter; one must also be an impartial and faithful judge. (Duhem 1906, 218)

Training oneself to be an "impartial and faithful judge" is easier said than done. It is often correctly pointed out that scientists can dogmatically cling on to their theories even in the face of contrary evidence. Max Planck said, "A new scientific truth does not triumph by convincing its opponents and making them see the light, but rather because its opponents eventually die, and a new generation grows up that is familiar with it" (Burchfield 1975, 165). This may be an overly cynical view and a little too sweeping. As we saw with the age of the Earth, scientists can and do change their views in the face of persuasive new evidence. But the deep reluctance to reject old beliefs and theories and accept new ones may well correlate with the amount of time that a given scientist has vested in using theories to reach, as Tolstoy said, "conclusions of which they are proud, which they have taught to others, and on which they have built their lives." One can understand why those scientists who, like Kelvin with the age of the Earth, had invested their entire lives within the framework of an old paradigm, acquired their prestige from it, and found it helped bolster other beliefs not directly related to it, may have been reluctant to relinquish it. It is possible that Kelvin succumbed to this temptation in his efforts to arrive at ages of the Earth that were too short for evolution to work.

Thomas Kuhn analogizes that adopting a new theory as a framework for scientific investigation is like retooling a factory. It may well result in a better product but it is time consuming and expensive and the process is only undertaken if one is forced to do so (Kuhn 1970, 76). Younger scientists or those who go into a new field of research are the ones who are more likely to adopt new theories since they have less vested interest in the old ones, and because new theories offer many more opportunities to make their mark with exciting new discoveries than old theories that have been well mined.

Darwin was well aware of this danger of self-deception because it is said that when working on his theory of evolution by natural selection, he knew that it would be controversial and that he would have to make the strongest case in its favor because people would look carefully for any piece of evidence that seemed to discredit it. Hence he was careful to quickly jot down any fact he encountered that seemed to *contradict* his theory in order to make sure he dealt with it properly, because he was aware that inconvenient facts tend to slip from one's mind. Ironically, his cataloguing of all the possible objections to his theory has been "quote-mined" by his critics as admissions of weakness. Although Darwin provided arguments as to why each objection was not valid, his critics tend to mention only the problems and ignore his counterarguments.

But not all scientists are as conscientious as Darwin, and modern day pressures to publish a lot and publish quickly can have the deleterious effect of causing some of them to not look carefully for problems, overlook them when they do appear, or too quickly dismiss them when they do register in their consciousness. The structure that scientists have set up, where their research findings are first subject to peer review by a small group of anonymous experts prior to publishing, and after publishing are open to further review by anyone, helps to reduce, though it can never eliminate, the biases of individual scientists. But here too there is popular misunderstanding about what peer review involves and what assurances it provides.

Before a paper is published in a reputable journal, the editors send the manuscript to other scientists familiar with the field to review the work and then, based on the feedback received, the editors decide whether to publish or not, or publish after revisions. In order to encourage frank reviews and reduce the effect of biases, this process of peer review is conducted anonymously in that the author of the paper does not know the identity of the reviewers. Less commonly, the author's identity and institutional affiliation are stripped from the paper before it is sent to the reviewers, so that only the editor knows the identities of the author and reviewers. But it is important to realize that a decision to publish does not mean that the reviewers and editors are certifying that the work is necessarily correct. Reviewers are not required to replicate the work and confirm the results. What they are expected to examine is whether the work says anything new and worthwhile, conforms to the standards in the field, properly cites other work, follows standard protocols, is internally consistent, has no obvious errors, makes sense, is essentially sound, takes into account implications for related areas of knowledge, and lays out all the information that would be necessary if someone else were to try and replicate the work. (This book went through this process.)

This last point is important because replication is not done for every paper and so it is possible, even likely, that there are many peer-reviewed papers in the literature that have results and conclusions that are incorrect. But those papers are usually those that do not have significant implications for the field and have little or no impact on the work of the community of scientists. Those errors lie dormant because there is little incentive in the structure of science for scientists to try to replicate other people's work just for its own sake and journals are loath to publish such replications. It is when a paper claims an important result or has a significant impact on other areas of science that it gets examined more closely by the larger community. In order to build upon that work, other scientists will often first try to replicate the results of the original paper that sparked their interest in order to make sure that they are at the correct starting point, and it is this process that usually turns up errors or, in rare cases, deliberate fraud. In other cases, researchers will take the results in the original work as a given (because scientists tend to trust the work of other scientists) but find that it causes problems when used in their own work and this discrepancy will trigger suspicions that something is wrong somewhere and result in a careful examination of all the possible contributory factors until the source of the error is unearthed. This process can take years, even decades.

Unfortunately the way that the media covers science suggests that such reversals are more significant than they really are. Scientific research is usually narrowly framed, with careful conditions imposed on the experimental conditions in order to better control the process to enable the drawing of meaningful conclusions about causal connections. Saying that "A causes B" implies that other possible factors C, D, E, . . . do not play any significant causal role. But it is not easy to identify all these other factors, weigh their contributions, and control for them. This is why scientific papers are usually *conditionally* phrased, couched with caveats about factors that were excluded in the present work but may impact the results and are thus left for future investigations. Most science proceeds in this incremental way, by pointing out future areas of study that would add to the piecemeal incorporation of new effects until a more comprehensive picture is obtained.

But the reporting of scientific results by the mass media leaves little room for such subtleties. It tends to ignore the caveats and takes the results as far more general, sweeping, and certain than the original authors might have claimed. Then when subsequent research finds that the original results do not hold up due to other factors playing a more significant role than originally suspected, the media again report this dramatically, as a reversal of thought by "science," when what it actually reveals is the natural incremental process by which ideas evolve and are refined until the scientific community reaches a consensus judgment that the phenomenon is well understood and that further investigation is unlikely to turn up anything new. This is how, for example, the age of the Earth ended up with the current widely accepted value of 4.54 billion years. The swings in earlier estimates were not due to errors or fraud but were due to the normal working out of the scientific process as new information and theories emerged.

Health, nutritional, psychological, and sociological research is particularly prone to this kind of media distortion because there seems to be a vast audience for breathless reporting about the latest medical breakthrough or wonder drug or miracle food or diet or explanations of people's behavior. There seems to be an insatiable market for a stream of announcements concerning scientific research about a new diet that makes people healthier or lose weight or concludes that something dramatically increases or decreases the chances of getting cancer or heart disease or stroke, or that something that in the past that was thought to be the cause of such feared ailments is now thought to be safe. The examples are too numerous and well known to be worth listing but the cumulative impact of such highly publicized repeated reversals is to cause some members of the public to be cynical about science in general even as they eagerly lap up the next so-called breakthrough. Unfortunately some scientists, as well as the institutions they work for and even the journals they publish in, share some responsibility for creating this misleading picture. The pressure to publish and get grants is now so great that in order to garner more attention, the results in a paper are sometimes exaggerated in institutional press releases (that the scientists usually had little hand in creating) to seem more dramatic than the way that they are presented in the actual paper, and these press releases, rather than the papers themselves, often form the basis of mass media reports, since the research papers themselves are often very dense and hard for laypersons to understand.

The resulting image of the fickleness of science is something that critics of science have seized upon to suggest that scientific conclusions are untrustworthy and just a species of opinion, one among many, and carry no greater weight than the speculations of others. The cartoon by Joe Heller (Figure 7.1) illustrates the problem.

But the cartoon scientist's assertion that science proves incontrovertible facts is also problematic, an example of the kind of folklore about the nature of science that is popularly held and that this book seeks to counter. It leaves science vulnerable to being undermined by those who have some knowledge of the history,

Figure 7.1 Credit: Joe Heller/Heller Syndication

philosophy, and epistemology of science and use it to advance their anti-science agendas by pointing to "facts" of the past that are no longer accepted as such. The scientific community has to counter the idea that scientific conclusions are just a species of opinion, without going to the other extreme and ascribing to them a level of infallibility that cannot be sustained.

The critics of science sometimes go even further and assert that the community of scientists is engaged in conspiracies to hide the truth from the general public. One sees this claim made by opponents of the theory of evolution who allege that scientists promote it because they have a secret anti-religious agenda. Some who deny that global warming is real or that human activity is a driver of climate change charge that scientists are manufacturing a fake crisis in order to obtain research grants. But the large numbers of scientists involved and the open way in which science is done makes these charges of cabals with secret agendas preposterous.

The community of scientists is well aware that an individual work is rarely the last word on any topic. Science relies upon the *collective* judgment of the community, in which individual biases get diluted, to arrive at conclusions that carry more weight than the judgment of a single member. This is similar to the way that legal systems are set up so that major decisions are made by a jury or a panel of judges rather than by a single person. The determining element in deciding whether a scientific claim is true is not that of an individual study or scientist, however reputable or famous that person may be, but the *reasoned judgment of the scientific community as a whole*, and this takes time to emerge. While there may be a period in which individual scientists or groups of them argue over the merits of competing theories, this state of uncertainty eventually gives way to a broad consensus, though sometimes the process can take a very long time, as we saw with the age of the Earth. As Duhem states:

> In any event this state of indecision does not last forever. The day arrives when good sense comes out so clearly in favor of one of the two sides that *the other side gives up the struggle even though pure logic would not forbid its continuation*. . . Since logic does not determine with strict precision the time when an inadequate hypothesis should give way to a more fruitful assumption, and since recognizing this moment belongs to good sense, physicists may hasten this judgment and increase the rapidity of scientific progress by trying consciously to make good sense within themselves more lucid and more vigilant. (Duhem 1906, 218, my emphasis)

As Duhem says and will be elaborated on in the next chapter, although pure logic cannot prove that one side or the other is correct or that one theory or another is the true one, by having different researchers subject theories to rigorous tests over and over again, eventually the weight of evidence tips the scales decisively in favor of one and the community of scholars that work in that field will arrive at a consensus. They do so using the rules of *scientific* logic in which the standard is not mathematical proof but is more similar to what is used in civil legal cases that require a *preponderance of evidence*, though depending on the scientific context, sometimes the more stringent standard used in criminal cases of *beyond reasonable doubt* is adopted. (This will be discussed further in chapter 14.)

Once a consensus emerges within a community of scholars or experts working in an area that there is a preponderance of evidence in favor of some scientific proposition or theory, then that proposition or theory is taken as being presumptively true. This system necessarily puts judgments about the truth and falsity of theories in the hands of specialists working in narrow fields of knowledge, which can seem undemocratic and cause concern because it implies that the validation of knowledge is delegated to a few scientists and that they are the modern day equivalents of ancient priesthoods, the sole possessors of knowledge. But what saves science from the charge of elitism or being controlled by secret cabals is that this "priesthood" is open to anyone who wishes to enter it, and its work and conclusions are open to scrutiny by anyone who wishes to do so. It is not the fault of science that nowadays it requires a long period of apprenticeship to achieve the requisite understanding in any area of research in order to properly evaluate it.

It should be borne in mind that the strength of the expert consensus is dependent not just on the diversity of the interlocking disciplines that are represented but also on the diversity of the practitioners who contribute to it. The paths followed by scientific inquiry are contingent upon the questions that are posed and those questions are influenced by the backgrounds of the people who ask them. It is undoubtedly true that modern science is largely the product of men from the western and developed world and recent feminist and post-colonial scholarship on the nature of science has pointed out the need for more diverse participation if the consensus views obtained are to be truly robust (Lederman and Bartsch 2001; Keller and Longino 1996; Harding 2006).

We should of course not blindly accept the consensus conclusions of experts however diverse they may be. But a reflexive mindless skepticism of expert opinion of the sort shown in the Heller cartoon in Figure 7.1 is far more dangerous. A long time ago Bertrand Russell said that while nonexperts should *defer* to the opinions of experts, this does not require that they uncritically accept their conclusions. Nor should they adopt a stance of "heroic skepticism" that doubts everything. He provided useful rules of thumb for how nonscientists might respond to the conclusions of scientists when they themselves do not have the expertise to carry out an independent evaluation.

I am prepared to admit the ordinary beliefs of common sense, in practice if not in theory. I am prepared to admit any well-established result of science, not as certainly true, but as sufficiently probable to afford a basis for rational action. If it is announced that there is to be an eclipse of the moon on such-and-such a date, I think it worthwhile to look and see whether it is taking place.

...

There are matters about which those who have investigated them are agreed; the dates of eclipses may serve as an illustration. There are other matters about which experts are not agreed. Even when the experts all agree, they may well be mistaken. Einstein's view as to the magnitude of the deflection of light by gravitation would have been rejected by all experts twenty years ago, yet it proved to be right. Nevertheless the opinion of experts, when it is unanimous, must be accepted by non-experts as more likely to be right than the opposite opinion. The scepticism that I advocate amounts only to this: (1) that when the experts are agreed, the opposite opinion cannot be held to be certain; (2) that when they are not agreed, no opinion can be regarded as certain by a non-expert; and (3) that when they all hold that no sufficient grounds for a positive opinion exist, the ordinary man would do well to suspend his judgment.

These propositions may seem mild, yet, if accepted, they would absolutely revolutionise human life. (Russell 1928, 2–3)

Duhem refers somewhat vaguely to the "good sense" of the scientific community eventually winning out to arrive at consensus conclusions that transcend those of individual scientists. As we saw with the age of the Earth, and as I will discuss again later, arriving at agreement among experts can sometimes take a long time to reach and even then, minority viewpoints can persist indefinitely.

The following sections of this book will elaborate on how this good sense comes about and enables us to be extremely confident of consensus scientific conclusions even though we cannot guarantee certainty. But this deeper understanding of what constitutes good sense is of value not just in science. It can vastly improve our ability to make rational judgment in every aspect of our daily lives and thus provides immense practical benefits.

PART THREE

SCIENCE AND TRUE KNOWLEDGE

8

A brief history on the search
for true knowledge

[The history of the search for true knowledge shows how scientific consensus conclusions have changed from being thought of as unchanging and infallible to now being considered as just *provisionally* true, the best we have at any moment. Establishing the validity of scientific propositions has become so difficult that it is now the preserve of a few specialists who have the time, resources, and expertise to carry out the required investigations.]

An important issue in any area of knowledge is being able to identify what is true and what is false. The search for what is true is the Holy Grail of epistemology, because those things that are true are believed to be of lasting value, while those that are false are ephemeral, often a waste of time, and at worst harmful and dangerous. Science has come to be associated with true knowledge and how this identification came about was discussed in the essay *The Demise of the Demarcation Problem* by Larry Laudan (Laudan 1983).

In trying to understand why science has become so closely identified with true knowledge, we can begin by going at least as far back as Aristotle (384–322 B.C.E.). He tried to make a clear distinction between those things that we feel we know for certain and are thus unchanging, and those that are subject to change. These two categories were variously contrasted as knowledge versus opinion, reality versus appearance, or truth versus error. Aristotle made the crucial identification that *true knowledge consists of scientific knowledge*, and this close association has persisted through the ages. It also made the ability to distinguish between scientific knowledge and other forms of knowledge, now known as the *demarcation problem*, into an important question since finding suitable demarcation criteria would enable us to distinguish truth from error and knowledge from opinion.

Aristotle said that the first characteristic of scientific knowledge was that it was based on foundations that were *certain* and thus *infallible*. Since he identified scientific knowledge with true knowledge, it followed that scientific knowledge had to be unchanging because how could something that was true ever become false? The second characteristic of scientific (and hence true) knowledge that distinguished it from other forms of knowledge was that it consisted of not just "know-how" but also of "know-why." Know-how knowledge is closely related to what we would now call technology and was considered to be the domain of craftsmen and engineers. Such people can (and do) successfully build boats, bridges, roads, houses, machines, and all manner of valuable and important things, often without needing a deep understanding of the underlying scientific theories and principles. The electrician I call to identify and fix problems in my house has plenty of know-how and does his work quickly and efficiently without having to understand, or even know about, Maxwell's laws of electromagnetism (the know-why) that scientists believe underlie all electromagnetic phenomena, whereas any scientist would claim that the latter was essential for really *understanding* the nature of electricity. It is this know-why knowledge, after all, that explains why the know-how knowledge works so well.

It was thought by Aristotle that it was this know-why element that had certain foundations and made us confident that scientific knowledge was true. As an example, even if one finds that all the wood one has encountered floats in water, this does not mean that the proposition that *all* wood will *always* float in water can be confidently asserted to be true, since it is conceivable that some new piece of wood might turn up that sinks (another example of the problem of induction). If one found such a piece of wood, it would be surprising but just an oddity, a black swan–like exception to a rule due perhaps to the type of tree from which the wood came from, and not as a revolutionary discovery.

But the scientific principle that all objects with a lower *density* than water will float while those with a higher density will sink seems to be on a much firmer footing, since that know-why knowledge penetrates to the core of the phenomenon of sinking and floating and gets at its root cause. If something with a lower density than water were found to sink or something with a higher density could be made to float, it would not be just a curiosity but would constitute a radical challenge to a basic scientific principle and result in investigations as to why it happened. For example, we know that it is possible to make a steel needle, despite its higher density, float on the surface of water. Explaining that exception requires invoking the concept of surface tension.

The importance of know-why for something to be considered science was the reason that Ptolemaic and early Copernican astronomy were not considered scientific during their time, even though they made highly accurate predictions of planetary motions. Their work was not based on an understanding of the laws that governed the motion of objects but on purely empirical correlations, and thus lacked know-why. If, for example, a new planet were to have been discovered, existing knowledge would not have been of much help in predicting its motion, even though astronomers would have expected it to have regular behavior like the other planets. That would have to wait until they had observed its behavior over some time to allow them to collect sufficient data to infer how it would move in the future. Hence astronomy was considered to be merely know-how and astronomers to be a species of craftsmen.

The work of Johannes Kepler (1571–1630 C.E.) and his postulation of a set of three laws that planets orbiting the Sun obeyed was a major step toward a deeper understanding of planetary motion. Note that his laws were *inductively* arrived at and were generalizations that were possible because of the careful observations of Tycho Brahe (1546–1601 C.E.) that corrected some older incorrect astronomical data. So although these laws constituted an advance from prior purely correlational understandings, Kepler's laws did not quite reach the level of know-why that would make them truly scientific. It took the arrival of Isaac Newton (1643–1727 C.E.) and his laws of motion and gravity to provide the underlying principles that governed the motion of planets. Newton's laws not only explained the existing extensive body of data on planetary motions and Kepler's laws, they were also able to *predict* the motion of any newly discovered planet. In 1846, they even led to the prediction of the existence of an entirely new planet, Neptune, and where it would be located in the solar system. Newton's theories provided the know-why that shifted astronomy into the realm of science.

As a consequence of the appreciation that know-why knowledge has widespread and more lasting value, science now largely deals with abstract laws, principles, causal relationships, and logical arguments. Empirical data are still essential of course, but mainly as a means of testing and validating those ideas. Many of these basic ideas are somewhat removed from *direct* empirical test and thus determining if they are true requires considerable effort. For example, I can easily determine if the pen lying on my desk will float or sink in water by just dropping it in a bucket. But establishing the truth of a scientific proposition, say about the role that relative densities play in sinking and floating, is not that easy and often requires sophisticated apparatus that is not readily available to everyone. As a result, this kind of know-why knowledge has shifted from being discoverable by anyone to becoming the preserve of the few people who have the time and other resources to do such investigations. The rest of us have become consumers of that information and have to rely on scientists to tell us what is true or not true. This shift has resulted in the general public becoming increasingly less aware of how science actually works.

The problem has become more acute as the search for knowledge has become more esoteric, more removed from everyday experience, requiring highly sophisticated technology to investigate, and as scientific communications largely shifted from books aimed at the general public to articles aimed at other scientists and published in specialized journals using language that is often unintelligible without extensive training in that field. The mathematical sciences (such as astronomy) had already become opaque to the general educated public even in early modern times and Newton's *Principia Mathematica* was likely read and understood by just a handful of his contemporaries.

Aristotle's idea of scientific infallibility, that the knowledge generated by science should be true and unchanging, was widely held for a long time. It received great support from Newton's three laws of motion and the law of gravity because those laws, while being so simple to state and understand that they are now taught even to middle school students, were believed to be universal in their application, used to explain the falling of apples from a tree as well as the motion of planets around the Sun. And they were immensely powerful. They could be used to solve difficult problems and make detailed predictions of behavior and indeed are still widely used and indispensable in everyday life. While anything other than the simplest configurations require sophisticated and detailed mathematics to obtain even approximate solutions, these were seen as technical difficulties that could be addressed by improvements in problem-solving techniques. Newtonian mechanics was seen as consisting of rock-solid truths. It seemed like we had arrived at true knowledge and it was science that enabled us to do so.

But this belief in the infallibility of scientific knowledge suffered a series of blows in the late nineteenth and early twentieth centuries that saw the repeated overthrow of seemingly well-established scientific theories to be replaced with new ones. Even the venerable Newtonian mechanics, long thought to be unchallengeable, was a casualty of this development. Newton's laws of mechanics were superseded by the special theory of relativity and his theory of gravitation was similarly eclipsed by the general theory of relativity. It became clear that Newtonian mechanics failed when dealing with phenomena that lay outside the range of our immediate experience, the worlds of the very small, very large, or very fast. Thus Aristotle's idea that scientific truths were infallible, universal, and timeless fell by the wayside, to be replaced with the idea that they were *fallible* and *provisional* truths, the best we had at the current time, that could be assumed to hold only until something better came along. But this new limited vision of science was one that was accepted mostly by philosophers. For many practicing scientists, the goal of finding ultimate, unchanging truths of science remained paramount.

One form that this commitment to truth takes is that it requires scientists to be truthful when reporting the results of their work because others depend upon it. The whole structure of scientific knowledge is created collaboratively, each person using and building upon the work of others. Because scientific knowledge is so interdependent, falsity in one area can do serious damage elsewhere, similar to the way that shoddy construction in even one small area of a building can threaten the entire structure. This does not mean that scientists are necessarily more truthful as *persons*. But it does mean that a structure has evolved where errors and falsehoods, especially if the results are significant and impact the work of others, cannot lie undiscovered for too long. Being dishonest is not a good career strategy for a scientist because there is a reasonably good chance that one will be found out. Scientists are not usually suspicious of the work of other scientists and do not reflexively check their work because it is not always feasible to independently verify every claim that others make. But the interdependence of knowledge means that an error or deliberate falsehood in one area will have consequences in other areas, and there is a chance that it will eventually be detected because people who use that knowledge will encounter inexplicable results. When the discrepancy is investigated in detail, the cause will eventually be traced back to the source. This is almost always how scientific errors and frauds are discovered.

As a minor example, when beginning my own research career as a graduate student in theoretical physics, our group started working on a problem where the application by others of a basic theory to a nuclear reaction produced results that did not agree with experiment. This discrepancy had been a source of concern in our subfield for some time because it seemed unlikely that the theory was false, which was why we chose to investigate it. However, when we did our own calculations using the same theory, we did find agreement with experiment, unlike our theoretical predecessors, and in investigating the discrepancy between the two theoretical calculations, we figured out that the earlier calculations contained an error. Thus the problem was satisfactorily resolved within the framework of the existing theory. Similarly, a few years later I published a paper with an error of my own and it was detected by others after a lapse of time by exactly the same process, because others had got different results than I did and had managed to diagnose my error. In both these cases, there was no dishonesty anywhere. The problems were due to the errors that can arise in any enterprise undertaken by fallible humans.

It is because of this kind of interdependence that science is largely, although not invariably, self-correcting. This is also why in academia, where the search for true knowledge is the prime mission, people who *knowingly* publish or otherwise propagate falsehoods or commit many errors suffer serious harm to their reputations and are either marginalized or drummed out of the profession. Some recent cases of misconduct include those of Jan Hendrik Schon's claims of manufacturing tiny, molecular-scale transistors (Chang 2004), the human cloning claims of Woo Suk Hwang (Cyranoski 2006), and Marc Hauser's results on cognition and morality in monkeys (Gross 2011). So accurately reporting honestly obtained data and making true statements about one's work is a prime requirement in science.

But there is another, more philosophically elusive search for truth that is also important and constitutes a collective enterprise of the scientific community as a whole. That consists of the drive to determine the truth of scientific *theories*. It matters greatly whether the special theory of relativity is true or not or whether certain classes of chemicals are carcinogens or not. To get those things wrong can have serious consequences extending far beyond any individual scientist. But it is important to realize that when it comes to theories, truth is *always a provisional inference made on the basis of evidence*, similar to the verdict arrived at in a legal case. And just as a legal judgment can be overturned on the basis of new evidence, so can scientific "truths" be similarly overturned, thus eliminating the notion of infallibility and reinforcing the idea that scientific truths are not eternal.

So given the primacy of scientific principles and laws in epistemology, and since the discovery of even provisional truths is to be always preferred over falsehood, an elaborate structure has grown around how to better establish the provisional truth and falsity of *scientific propositions*. This often requires the construction of expensive and specialized equipment to determine the empirical facts relating to those propositions, and extensive long-term study of esoteric subjects to relate the propositions to the data. For example, the long and expensive search for that elusive particle known as the Higgs boson that required the construction of a massive accelerator is symbolic of the extreme levels that this kind of effort can take. The Large Hadron Collider took over ten years to build, cost over $10 billion, and required the work of thousands of scientists, engineers, technicians, and other workers from all over the world. As a result of all that effort it is now confidently asserted that the Higgs particle exists, but only a handful of people really know what the evidence is in favor of it or even what its discovery represents. And even those few are dependent on their many team members reporting their part of the investigations accurately. The rest of us just have to take their word for it.

The confirmation of the existence of the Higgs particle has had no direct impact on the lives of people. It was searched for because it was the prediction of a fundamental theory of physics and played a vital role in how the theory worked. The discovery of the Higgs meant that the theory passed an important test. Note that since the theory had long been thought to be true anyway, the discovery did not really change anything. But testing the predictions of theories constitutes an important facet of the search for true knowledge, always bearing in mind that the term "true" is being used provisionally.

So to summarize, science was once thought to work so well because the knowledge generated by it was infallible and eternally true. We now know that that cannot be the case and that scientific knowledge is only *provisionally* true, the best we can do at any given time. But science still works so exceedingly well in solving so many of the most urgent problems that we as individuals and society face that trying to unearth the reasons for its tremendous success has become an important and interesting question for research.

9

The role of doubt and faith in science

[It is important for *all of us* to understand the epistemology of science in order to better counter those who selectively use such knowledge to advance dangerous anti-science agendas, such as arguing that science is just another species of opinion and requires faith. While doubt is always present in science, even in the absence of absolute certainty we can still have a high degree of confidence in scientific consensus judgments.]

Why *does* science work so well?

Some people might think that science is so successful because the *methods of science* enable us to distinguish conclusively at any time between those scientific theories that are true and those that are false. Eminent physicists from a century ago such as Albert Einstein, Niels Bohr, Erwin Schrödinger, and Werner Heisenberg were deeply interested in this question of truth and felt that the philosophy of science could provide clarity and understanding. For example, here is Einstein in 1944 in a letter to Robert Thornton, a young philosopher of science who had just finished his PhD and who had written to him seeking support for his efforts to introduce philosophy of science into his modern physics course.

> I fully agree with you about the significance and educational value of methodology as well as history and philosophy of science. So many people today—and even professional scientists—seem to me like somebody who has seen thousands of trees but has never seen a forest. A knowledge of the historic and philosophical background gives that kind of independence from prejudices of his generation from which most scientists are suffering. This independence created by philosophical insight is—in my opinion—the mark of distinction between a mere artisan or specialist and a real seeker after truth. (Howard 2017)

This was no casual statement tossed out by Einstein turning to philosophy in his later years but reflected a long-standing belief. Much earlier in 1916, just a year after his groundbreaking paper on general relativity, Einstein had written on the occasion of the death of Ernst Mach:

> How does it happen that a properly endowed natural scientist comes to concern himself with epistemology? Is there no more valuable work in his specialty? I hear many of my colleagues saying, and I sense it from many more,

that they feel this way. I cannot share this sentiment. When I think about the ablest students whom I have encountered in my teaching, that is, those who distinguish themselves by their independence of judgment and not merely their quick-wittedness, I can affirm that they had a vigorous interest in epistemology. They happily began discussions about the goals and methods of science, and they showed unequivocally, through their tenacity in defending their views, that the subject seemed important to them. Indeed, one should not be surprised at this. (Howard 2017)

Philosophers, historians, and sociologists of science, most of whom are supporters of science, have tried to understand and elucidate why science is so successful but despite looking long and hard, have not been successful in finding any *methodological* keys that prove that its theories are true. In fact their scholarship seems at first glance to undermine the very foundations of science by taking away the sense of certainty about the scientific knowledge we depend upon in our everyday lives. Beginning in the latter half of the twentieth century, some of the knowledge generated in those fields has been selectively seized upon and misleadingly used by those who seek to discredit science and undermine its credibility.

As a result, some contemporary scientists have shifted from the ignorance and indifference to epistemology that Einstein lamented to being actively hostile and openly contemptuous of the field, and have lumped those scholars together with the enemies of science (Theocharis and Psimopoulos 1987). These scientists have tended to ignore or disparage the results of the scholarly efforts of philosophers, historians, and sociologists of science since those do not provide unqualified support for the scientists' own simplified view of the reasons for science's success, that it works so well because it must be true or approaching truth.

Many scientists feel that they understand why science works and are confident that it leads to true knowledge even if they cannot explain the reasons for this confidence. The fact that it works so well seems to be a sufficient argument for them. If pressed, they might appeal to the idea of falsification, that contradiction with experiment can weed out false theories leaving only those that are unfalsified as potentially true. That falsification has been found to be untenable as a mechanism for adjudicating the truth of theories (as will be discussed in more detail in the next chapter) is casually dismissed. These scientists tend to take the attitude that if philosophers cannot find the line of reasoning that shows that science works so well because it is true or approaching truth, that is not a problem for scientists but for philosophers and they should work harder at it. The philosophical dogs may bark, but the scientific caravan moves on.

While philosophy does not *lead* to new scientific discoveries, this dismissal of the philosophy and epistemology of science is a mistake since these fields can help us understand the nature of science and better explain it to the general public. Since supporters of science are often unfamiliar with the philosophical arguments used by the more sophisticated enemies of science, they tend to not be able to combat their critics effectively. Understanding at a deeper level the issues that philosophers and historians of science are raising can help us better refute the arguments of those who selectively use that knowledge to advance harmful anti-science agendas.

To start with, we must first acknowledge that there is *always an element of doubt* about the truth of *any* scientific proposition, theory, conclusion, or even fact. If you ask any serious scientist whether she is 100% *certain* of the truth of her theories or even of her data, she will reply "no" because scientific history alone, with its many repeated reversals of knowledge, undermines that supposition. Scientists will always allow for a small window of doubt that they might be wrong. After all, even if there do exist unique and unchanging truths about the universe, and even if by some chance we stumble upon them, how would we *know* that we have done so? No gongs will sound or cymbals clash to herald that success. The fact that a scientific law has withstood challenge for a long time is no guarantee of its truth and permanence, as the problem of induction keeps reminding us. Newton's laws are a prime example of something that was firmly believed to be true and universal for over two hundred years but was then found to be not so.

Scientists are accustomed to being in a state of doubt and are not paralyzed by it. Indeed such doubts are the drivers of scientific progress. As Richard Feynman said:

> Scientists, therefore, are used to dealing with doubt and uncertainty. All scientific knowledge is uncertain. This experience with doubt and uncertainty is important. I believe that it is of great value, and one that extended beyond the sciences. I believe that to solve any problem that has never been solved before, you have to leave the door to the unknown ajar. You have to permit the possibility that you do not have it exactly right. . . . And it is of paramount importance, in order to make progress, that we recognize this ignorance and this doubt. Because we have the doubt, we then propose looking in new directions for new ideas . . . . So what we call scientific knowledge today is a body of statements of varying degrees of certainty. Some of them are most unsure; some of them are nearly sure; but none is absolutely certain. Scientists are used to this. . . Doubt is clearly a value in the sciences. (Feynman 1998, 26–28)

Some people with dubious agendas in particular areas have tried to exploit this concession by trying to pry wide open this tiny window of doubt to create an image that science operates in a state of massive uncertainty. Of course they cannot do this for all of science since modern life depends on the accuracy and reliability of science and to not have confidence in anything is to be paralyzed by indecision and look ridiculous. What they do is focus their attention on those specific areas of science that they have a vested interest in in order to prevent any action that harms their interests, or to preserve beliefs that they hold dear but cannot justify.

The first category of critics consists of those protecting their own economic interests. For example, a coterie of people, supported largely by the fossil fuel industry, tries to deny that climate change is happening or that if it is, it is not caused by human activity. They do this in order to prevent any governmental actions at local, national, and international levels that might impinge on their economic interests, and they even try to dismantle existing regulations that inhibit their freedom to act as they wish. These groups have mounted a concerted attack on the scientific climate models that are predicting catastrophic consequences if we do not reverse, or even just limit, the current levels of human-generated greenhouse gas emissions that are threatening to raise the average global temperatures to a level that will lead to, and indeed are already displaying, the melting of the polar ice caps, rising ocean levels, and erratic and extreme weather patterns. As Naomi Oreskes and Erik M. Conway point out in their book *Merchants of Doubt*, for these groups "the goal was to fight science with science—or at least with the gaps and uncertainties in existing science, and with scientific research that could be used to deflect attention from the main event" (Oreskes and Conway 2010, 13).

The second category of science critics consists of those who are alarmed at the threat it poses to cherished beliefs. This group comprises astrologers, homeopaths, psychics, faith healers, purveyors of dubious health remedies, and the like, plus assorted religious groups. Influential groups in the United States that fall into this last category are the creationists who believe in an Earth that is only a few thousand years old. As we saw in chapter 6, they also fear that the theory of evolution by natural selection makes the existence of God unnecessary in understanding the diversity of species and undermines the idea that human beings are special creations whose appearance in our current form on the Earth was preordained.

Members of both categories realize that the scientific *evidence* against their positions and beliefs is overwhelming and they do not have sufficient counter-evidence to challenge it. So they take the tack of *arguing against science itself* in order to try and shake the belief that the conclusions derived using the evidence-based reasoning of science are the ones to be preferred when making important decisions. They selectively use scientific facts and certain elements about the nature of science to undermine any conclusions that go against their interests. They exploit the features of rational decision-making theory that says that in the case of uncertainty, your best outcome is generally to do nothing (Giere, Bickel, and Mauldin 2006). Most people, when faced with doubt or the choice between competing theories that they are not in a position to evaluate for themselves and with no compelling reason to favor any specific one, tend to favor the status quo and taking no action or choose the option that is easier or less expensive or less discomfiting. Since science is challenging the status quo when it comes to climate change or the various superstitions, the groups threatened by it have set as their goal to make people believe that the consensus views of the scientific community are just opinions, merely one set of views among many others of equal credibility and thus have no greater epistemic justification, so taking action based on them is unwarranted.

Both these groups selectively use science and the philosophy of science to undermine the credibility of science. They argue that since science is never certain about the truth of its theories, the consensus view of the vast majority of scientists on any issue is on the same epistemological footing as views put forward by any minority group of scientists, however tiny, or those on the fringes of science or even nonscientists. Some go further and assert that science is like religion in that both involve *acts of faith*, believing in ideas that are unprovable, and so scientific conclusions have a similar status to all the other belief systems that cannot be proven to be true.

They are wrong. Although the word "faith" is used in science, like so many other words what it means in science is quite different from how it is used in everyday life. While science no longer makes unequivocal claims to being true, "scientific faith" (if one wishes to use that phrase) is what is necessary for someone to decide that a theory is so well supported that it can be considered *effectively true* and not have to worry that it might be false. In other words, faith in science can be used to describe the transition from *almost* certain to certain.

I get on planes without a qualm. That is because I have "faith" that the entire spectrum of scientific theories such as the laws of hydrodynamics, gravity, mechanics, electromagnetism, thermodynamics, on so on that go into designing and building a plane and making it fly are all true and that they will remain true until the end of my journey. I do not concern myself at all with the possibility that these laws might fail partway through the flight and cause the plane to crash. But have all these theories been *proven* to be true? Are we 100% certain? No and no. There is always the possibility that at least one, or some of them, or even all of them, is false. But they have been so thoroughly investigated and tested that I can have an extremely high level of confidence in them. From there, it is but a small step to being certain about them, to treating them as true. Taking that small last step *without any concern of being wrong* is what constitutes the word "faith" when used in science. As physicist Seth Lloyd says, "Unlike mathematical theorems, scientific results can't be proved. They can only be tested again and again until only a fool would refuse to believe them" (Lloyd 2006, 55).

As a metaphor, consider a wide and raging river that one has to cross. What science does is build a bridge across that river that takes you to the other side. Almost the entire length of the bridge has been thoroughly tested and found to be solid and reliable, all but the last bit that is close enough to the other bank that one requires just one step on the last bit to go from where the well-tested part of the bridge ends to the other side. This last bit has been constructed in the same way as the well-tested part. As a result, one can take the last step with supreme confidence, with faith if you like to use that word, knowing that the principles used to make the small last segment are the same as the ones used to make the highly tested ones.

No other knowledge claims have such a solid basis for confidence. People who make important life decisions based on the words of astrologers, homeopaths, psychics, faith healers, and preachers are taking huge risks because those claims have not been subjected to anywhere near the same number, variety, and stringency of tests as scientific theories have been. These alternatives are the equivalent of creating a ramshackle structure across the river that has not passed any serious test. Some people might have faith that they can cross the river using such a bridge but the prudent person will realize that it is folly. To put one's trust in such claims is to invite disaster and many lives have been ruined by doing so, such as people making all manner of foolish decisions or even dying of curable ailments because they abstained from medical treatment in the belief that faith would cure them or their children.

Faith in science involves making the assumption that 99% confidence (the exact number does not matter as long as it is close to 100%) is close enough to certainty that we can with equanimity assume it to be 100%. On rare occasions, that 1% will turn out to be significant and we will be proven wrong and a new theory will emerge, and how and why that happens will be discussed later in the book. But while we cannot be 100% *certain* of the truth of our scientific theories, we can get close enough that we can act with 100% *confidence*, so that we can ignore the possibility of them being false as we go about our lives. As an analogy, we have more confidence that the Sun will rise tomorrow than that we will live to see it. But we have sufficient confidence in both that we can ignore the possibility of either one not coming to pass.

Of course, not all theories of science carry with them the same level of confidence. Especially in the medical and social sciences, scientific papers can be published even if they have just 80% confidence that the results were not arrived at by chance. But in areas of biology and medicine and psychology, practitioners are aware that they work in an environment that is harder ethically and practically to control experimentally, and that the results are necessarily less certain.

They have developed methods to enable them to tease out reasonably firm conclusions despite those limitations, but are also duly cautious (or should be) about making claims that are too strong (Jenkins 2004).

In order to appreciate the difference between the different kinds of faith used in science and other areas, we need to address the epistemological foundations of science directly and this is what subsequent chapters will seek to do, by addressing how science can work so well when its theoretical foundations seem to lack certainty. We will see that despite this limitation, there are excellent reasons to trust the consensus conclusions of the scientific community. When crossing the rivers of life, the scientific bridge is always to be preferred over the others.

10
The basic features of science

[This chapter addresses some popular misconceptions about the nature of scientific theories. It discusses why they cannot be proven true or false, that scientific revolutions always involve at least a *three*-cornered struggle involving at least two theories and experiment, and that experimental data are never sufficient to uniquely determine a theory. Although we have not been able to specify both necessary and sufficient criteria to distinguish science from nonscience, there do exist necessary conditions that any scientific theory and law must satisfy, and that is that they must be both *naturalistic* and *testable*.]

To sum up what has been discussed so far, science is *fallible* in that what we think is correct now may well be shown to be wrong in the future. Scientific knowledge is thus *provisional* and subject to change over time and as such there is every reason to think that science does not have, and will *never* have, definitive answers to all the questions. And even if there does exist a set of universal and eternal true laws out there just waiting to be discovered, there are no external indicators to tell us that we have found them. How would we know if we have stumbled upon them, since agreement of theory with experiment and observation over a long time is by itself no guarantee of truth, as the example of Newton's laws shows us? As stated earlier, the problem of induction limits what we can infer purely from data. However many times something has behaved in exactly the same way in the past, however many times a theory has passed even the most stringent tests, there is no *logical* reason to believe that it will do so the very next time.

All these considerations seem to undermine the idea of certainty in the knowledge that science produces and makes its undoubted success seem inexplicable. It is the use of *scientific logic* that makes science so powerful despite these seeming limitations. The rest of this book will discuss the nature of scientific logic and show how the knowledge structure that is built using a combination of that logic with empirical evidence enables us to arrive at conclusions that are remarkably robust and reliable, and worthy of the immense trust we place in them.

This is not merely an academic exercise but has immense practical consequences for everyone. It enhances people's ability to make rational decisions because *scientific logic is not used only in science*. In fact, most of us depend upon it *all the time*. The logic used by the scientific community to arrive at conclusions in areas that are highly esoteric consists of the same kind of reasoning that we often use to arrive at conclusions in our everyday lives. It would be impossible to go about our daily lives without using it and yet most people are unaware that they are doing so and thus use it idiosyncratically. It is scientific logic that enables most of us to confidently assert that zombies, vampires, mermaids, unicorns, werewolves, and a whole host of mythical characters do not exist despite their popularity in the folk literature. It is what enables us to have such confidence that the Sun will rise in the East tomorrow and that spring will follow winter. So when critics of science reject some of the conclusions of science because they are uncongenial for whatever reason, we should make them aware that *they are also rejecting our entire way of thinking about almost everything.*

The success of science has been so exceptional that it has spawned entire academic disciplines of the history, philosophy, and sociology of science to unearth the secret of its success despite its seeming epistemological weaknesses. In addition to the idea that scientific laws and theories are fallible and provisional, other consensus views of scholars in this field can be summarized in the following statements:

(a) Science does not *prove* theories to be *true.*

(b) Science does not *prove* theories to be *false.*

(c) Scientific revolutions are not triggered simply by the disagreement of a theory with experiment but are always the result of *three-cornered struggles* involving at least two theoretical frameworks and experiment.

(d) *All theories are underdetermined*, in that any set of data, however large, never uniquely determines a theory. Even if a theory perfectly explains a set of facts, *you can always postulate an alternative theory, indeed infinitely many theories, to cover that same set of facts.*

(e) Scientific theories must be *naturalistic* and *testable.*

Some of these statements will be more obvious than others, depending on the extent to which the reader is familiar with science and its epistemological literature, and some are more strongly held as consensus views than others. Each will be explored in more detail in subsequent sections of this chapter and while on the surface they may seem to undermine beliefs in the truth of scientific theories, we will see that we can still have immense confidence in those theories. Indeed the opposite view, that we can prove the truth and falsity of theories and that theories can be uniquely determined by data, while superficially seeming to strengthen belief in the power of science, actually undermine it, because critics can easily point to counterexamples that expose the weaknesses of those positions.

(a) Science does not *prove* theories to be *true*

What constitutes a proper test for a theory is somewhat problematic, both at practical and deeper philosophical levels. Take as an example the simple statement of Newton's second law of motion that states that the net force on a body is equal to the rate of change of momentum of that body. A common operational form of that statement is given by the equation $F = ma$, which says that if one measures the net force F acting on a body, that number will be equal to the number obtained by multiplying the mass of the body m by its acceleration a. For over two centuries after Newton proposed this law, until the arrival of the theories of relativity and quantum mechanics, it was thought that this law governed the behavior of *all* entities that had mass, and held true under *all* conditions. On what basis was it believed that this law was true?

The teaching of science at the precollege and even introductory college levels follows a fairly standard approach that tends to give the misleading impression that theories can be proven to be true and even how it is done. Scientific laws are taught in a *deductive* manner, going from the general law to particular cases, with little understanding of how the general rule was arrived at to begin with. A scientific law (say Newton's second law or the law of conservation of energy) is first enunciated with its terms carefully explained. It is asserted that this law is applicable in all situations or at least a wide array of them. Students are left with the vague impression that the laws were "discovered" (in much the same way that fossils are discovered) or by some form of revelation. Anecdotes, myths, and folklore surrounding the circumstances of the discovery are recounted to support this view. Newton and the falling apple inspiring his theory of gravity, and August Kekulé's daydream of a snake seizing its own tail leading him to his discovery of the cyclic structure of benzene, are particularly vivid and well-known examples. The way the law applies in a few specific carefully controlled situations that can be investigated experimentally is then elucidated. Students are then shown demonstrations or do experiments that seem to provide results in agreement with the law. The law is then believed to have been "confirmed" to be true, at least in those instances, and inductive reasoning is then invoked to suggest that it would always be found to be true if one invested sufficient time and effort to do more tests.

When the public is taught that this is the way that scientific knowledge is acquired and justified, it is not surprising that scientific reversals about what is considered true can cause concern. It seems to imply that the demonstrations or experiments that initially confirmed the law must have been incorrectly done, and this leads to skepticism toward either the competence or the integrity of the scientists who originally "discovered" the existence of the law and those who did the experiments that were believed to confirm its truth. This enables enemies of science to seize upon scientific reversals and hoaxes to try to discredit science by suggesting that its conclusions are unreliable. For example, young Earth creationists are taught as children specific examples of scientific reversals such as those involving Piltdown Man and Nebraska Man, and are trained to ask science supporters how science can be trusted if such hoaxes and mistakes were once believed to be major discoveries. They know that no supporter of science will be aware of the detailed circumstances behind every single discredited theory and thus can be discomfited by not having the knowledge at their fingertips. The critics will often raise biology examples when talking to physicists and physics examples with biologists. The better defense of science is to go beyond the details of specific cases and argue that *such reversals are the norm in science* and do not undermine its credibility. Such a response has far greater general utility than researching every single case of reversal but requires a more sophisticated knowledge of the epistemology of science.

A little thought will reveal that there are two problems with trying to *prove* the truth of the laws of laws of science by using confirmatory evidence. One is purely practical and arises from a basic characteristic of measurement. When an actual experiment is done (like measuring the height of someone using a tape measure), what we obtain are readings given by a tape or pointer or other display on a measuring instrument. There is *always* some uncertainty involved in extracting the value of the measured quantity from the reading. Part of it may be due to visually taking the reading, although this may be eliminated with the use of digital displays. The more difficult uncertainties to deal with are those that arise from the limitations of the measuring instruments themselves and the lack of perfect control over the conditions of the experiment. Hence we cannot know each quantity *exactly* and the best that we can do is to obtain a *range* of possible values for each measured quantity. It should be emphasized that this lack of exactness is not due to any failing of the experimenter or the measuring devices. Though both can influence the size of the range of uncertainties, they cannot make it zero. Heisenberg's Uncertainty Principle says this inability to measure with perfect precision is even a fundamental property of nature for certain quantities.

This uncertainty makes it almost impossible to say if two measured quantities are equal. For example, suppose we have two people who seem to be of the same height. Can we say that they are exactly equal? If we measure them with a tape measure and both come out to be 1.7 meters tall, does that mean they are the same height? Is it not possible that if we measured them more precisely with a more sophisticated measuring instrument, there could be very slight differences, say of the order of a centimeter? But however precisely we measure them, there is always the possibility that there exist even tinier differences that our measuring devices cannot distinguish, and this prevents us from ever definitely concluding that two people are *exactly* the same height.

Unfortunately, the teaching of science in schools tends to skirt the profoundly important truth about the nature of scientific inference embedded in this simple aspect of measurement. Students are given the impression that whatever reading the measuring instrument gives them is the exact value and tend to ignore the uncertainties inherent in the measurement. Thus they expect to find exact agreement for $F = ma$ (after all, this is called a "law" discovered by the scientific giant Newton and thus carries with it great prestige) and strive to get it, using just the pointer readings alone. Exact agreement will never happen in reality but any discrepancy is casually dismissed as being due to inadequacies of the equipment or the skills of the experimenter. It is believed that if the experiment were done correctly, exact agreement would be obtained.

The reality is that there will *always* be an element of uncertainty as to whether any numerical relationship such as $F = ma$ involving measured quantities is satisfied exactly or not, because each measured quantity is not known precisely, only to lie within some *range* of values. So while the relationship might be *approximately* satisfied within this range, we cannot be sure that it is *exactly* satisfied. It is after all conceivable that, as with the heights of people, the numbers would be found *not* to agree if the measurements could be done more precisely. It should be emphasized that this inability to get exact agreement is the *normal* situation in science, not the exception, and thus must be addressed when we are talking about scientific progress. If scientific laws are represented by numerical relationships, and if we cannot ever determine that the relationships are exactly satisfied due to unavoidable uncertainties in measurement, how can we say that a law was confirmed to be true by experiment?

A second problem with the confirmatory approach is even more serious. A law is not a statement about a relationship involving just *one* particular set of values for the relevant quantities, or even a few sets. It is assumed to be valid for *all* the possible experiments that can be carried out within its domain. Newton's law is not just about what will be the acceleration of one particular mass when one particular force acts on it. It is believed to be true for *all* masses and *all* forces. So even if we ignored the problem of uncertainties and could confirm the law for one case or a few or even many cases, how can we be sure that it is true for *all* cases? After all, any nontrivial theory will allow for predictions of results of an *infinite* number of experiments but we can only perform a finite number of them.

This is the problem of induction rearing its head again. No matter how many times a theory passes a test, we cannot be sure that there is not some new test out there, perhaps the very next one that we try, that will lead to failure. How many confirmatory results would be needed before a theory would be accepted as true? How can we arrive at any particular requisite number of experiments that is not purely arbitrary?

This is also a problem for attempts to use probabilistic arguments to quantify the *likelihood* of theories being true. While we can assign probabilities to the results of individual experiments that have a *finite* number or range of outcomes, we cannot do so for the *theories* that underlie them. For example, it has been stated with 99.99994% confidence that the results of the experiments at the Large Hadron Collider were due to the Higgs boson and not due to some other factors. This lends support to the idea that the Standard Model of particle physics that predicted the Higgs is correct. But that 99.99994% confidence level that what was seen was the Higgs cannot be extended to the probability of the truth of the Standard Model that predicted the existence of the Higgs boson. This is because we cannot go from the probability of the outcome of specific *events* predicted by a theory or hypothesis to the probability of the theory or hypothesis being true.

As philosopher of science Karl Popper argued, all attempts to go beyond intuitive and subjective judgments about the probability of a theory being true and try to assign a numerical value to it immediately run aground because the number of possible predictions of any nontrivial theory is infinite, while the number of experimental results is finite. Hence the probability of even our most trusted theories being true is effectively of measure zero.

> One [way] would be to ascribe to the hypothesis a certain probability—perhaps not a very precise one—on the basis of an estimate of the ratio of all the tests passed by it to all the tests which have not yet been attempted. But this way too leads nowhere. For this estimate can, as it happens, be computed with precision, and the result is always that the probability is zero. (Popper 1968a, 257)

Another philosopher of science, Rudolf Carnap, tried to use inductive logic to create a confirmatory approach to scientific laws but he too could not find a way out of the conclusion that the degree of confirmation of every universal law is always zero (Murzi n.d.).

These problems with the confirmatory approach are what make the search for proving a theory to be true, or even assigning probabilities as to its truth, a futile exercise. While scientists do make statements about the relative probabilities of different theories being true, these are *intuitive* judgments based on other grounds, such as perceptions of the quality of the evidence and reasoning that are made in support of the theory, and not on the basis of any quantitative estimates of probabilities. Assigning numerical probabilities to the truth of scientific theories turns out to be unworkable in practice and earlier generations of philosophers of science who had tried to do so abandoned their attempts. We are misled into thinking we should be able to do so because of our desire to quantify our intuitive judgments about the relative merits of competing theories.

This has not prevented more recent efforts by philosophers of science to try and revive the idea that numerical probabilities can be arrived at and play a role in evaluating the relative merits of theories, and what are known as Bayesian methods have been proposed to fill that need. For the purposes of this book it is not necessary to understand Bayes' theorem or go into the details of this approach. Its basic idea is that if we can assign an a priori probability to a proposition being true, additional evidence that comes along can be used to adjust that probability accordingly. The problem is that while the method can be used to find the probabilities of specific outcomes of an experiment, it fails when it comes to assigning probabilities for the truth of the *theories* that predicted those outcomes. So for example, we can use statistics to assign probabilities to the efficacy of a particular medical treatment and of specific drugs, but that does not help us to assign a probability that the theory on which the treatment or drug is based is true. Supplementary Materials A section has more information on why Bayesian statistics fails in its attempts to assign numerical probabilities for the truth of scientific theories.

But even conceding the highly debatable point that Bayesian or other probability models can be used to gauge the relative merits of theories, this approach suffers from the fatal flaw that there is no evidence that scientists actually *use* the ideas of probabilities, Bayesian or otherwise, in any concrete way when making judgments about the truth of theories or in comparing theories. In other words, it does not describe how an individual scientist or the scientific community as a whole goes about its work in comparing and evaluating *theories*, as opposed to evaluating the outcomes of specific experiments. While they may use the language of probabilities and speak of theories in which they have great confidence as having a high probability of being true and of highly speculative theories as having a low probability of being true, they make little effort to assign numerical values to those statements, leaving them as purely subjective impressions.

(b) Science does not *prove* theories to be *false*

Most scientists realize that it is impossible to *prove* that any scientific theories are true. Popper and Carnap also nailed the door shut to the idea that we could use probabilistic ideas to assign numerical values of likelihoods to theories or laws being true or to compare one theory with another to see which one was more true. But Popper opened another door with a new idea that seemed to enable objective judgments about theories. While trying to prove that a theory is true is defeated by the intrinsic uncertainty that accompanies all measurements and the fact that we can only do a finite number of tests out of the infinite number that are possible, what he said was that both those limitations go away if one's goal is to try and prove a theory to be *false*. This idea of *falsification* as an explanation of how science advances has gained a powerful hold on the minds of scientists and nonscientists alike because of its appealing simplicity and the fact that much of scientific folklore seems, at least on the surface, to support it.

As an example of how falsification works, we return to our example of Newton's second law of motion expressed as $F = ma$. As discussed earlier, if (say) the three measured quantities in $F = ma$ agree within the range of uncertainties, then all we can conclude is that the relation *may* hold exactly but there is also the possibility that it may not hold. But if the range of values of F lies *outside* the range of possible values of ma, then we can definitely conclude that the relationship is violated and that the law is false.

To make this idea concrete, suppose the measuring instruments used to test $F = ma$ give a reading of 100 for the value of m and a reading of 10 for the value of a. Suppose also that each of the measurements can be done with an accuracy of 1%. What that means is that the true value of m lies somewhere in the range 99 to 101 and the true value of a lies in the range 9.9 to 10.1. (For the sake of simplicity of argument I will ignore the fact that with statistical uncertainties, what we get are not absolute limits to the range but *probabilities* that the true value lies within the range.) Thus the smallest possible value of the product ma is 99 multiplied by 9.9 which gives 980, while the largest possible value of ma is 101 multiplied by 10.1 which gives 1020, rounding both numbers to the nearest integer. When we measure the value of F, it too will have a range of possible values. If that range overlaps with the range from 980 to 1020, then all we can conclude is that the relationship *may* have held although we cannot conclude that it has held exactly.

But Popper's insight was that we were looking for the wrong thing. Rather than trying to prove a theory right, a futile task, we should be trying to prove it wrong because such an occurrence would be unambiguous.

> These considerations suggest that not the *verifiability* but the *falsifiability* of a system is to be taken as a criterion of demarcation. In other words: I shall not require of a scientific system that it shall be capable of being singled out, once and for all, in a positive sense; but I shall require that its logical form shall be such that it can be singled out, by means of empirical tests, in a negative sense: *it must be possible for an empirical scientific system to be refuted by experience.* (Popper 1968a, 40–41, emphasis in original)

So if the range of possible measured values of F *does not overlap* with the range of 980 to 1020, then we can conclude that the theory is false, despite the uncertainty of each individual measurement that prevents us from knowing the value of each term precisely. Hence, rather than trying to show that a theory is true with all its attendant complications, it seems to be much more straightforward and unambiguous to focus on experiments that have the possibility of proving it false because to do that all we have to do is find a *single* experiment that *disagrees* with the predictions of the theory. Scientific theories and laws are believed to be statements about nature that are not capricious and brook no exceptions. As soon as the existence of even a single experiment that disagrees with a theory is conclusively demonstrated, then according to the falsification approach, that theory is deemed to be false and must be rejected.

Popper originally introduced this idea in the context of addressing a long-standing and vexing issue known as the *demarcation problem*, which was that it had not been possible to find ways to unambiguously distinguish science from nonscience even though, since Aristotle's time, the former had been associated with truth while the latter was considered mere opinion. The immense prestige accruing to science from its manifest successes had resulted in people claiming the imprimatur of science for all manner of theories in the hope that the same respect would rub off on them. Popper seemed particularly irked by claims of their respective supporters that Marxist theories of economic and political development, Freudian theories of psychoanalysis, and Jungian theories of psychology were also scientific. These groups would point to success after success for the predictions of their theories as evidence that they were true, or at least highly probable of being true, thus meriting their appropriation of the label of science, since science was identified with generating true knowledge.

But Popper felt that all these claims carried little weight because the theories were so flexible that almost any result could be accommodated within them and trumpeted as a success. He viewed such successes as so cheap as to be worthless and his antipathy toward those theories led him to search for demarcation criteria that would exclude them from claiming the prestige that came with being perceived as scientific. His falsification model was based on the idea that whether a theory was scientific or not depended on whether it subjected itself to serious and rigorous tests that *had the potential to show it to be false*. It is worth quoting Popper more fully on this question (Popper 1965, 36). He makes seven points (emphasis in original):

1. It is easy to obtain confirmations, or verifications, for nearly every theory— if we look for confirmations.
2. Confirmations should count only if they are the result of *risky predictions*; that is to say, if, unenlightened by the theory in question, we should have expected an event which was incompatible with the theory—an event which would have refuted the theory.
3. Every "good" scientific theory is a prohibition: it forbids certain things to happen. The more a theory forbids, the better it is.

4. A theory which is not refutable by any conceivable event is non-scientific. Irrefutability is not a virtue of a theory (as people often think) but a vice.
5. Every genuine *test* of a theory is an attempt to falsify it, or to refute it. Testability is falsifiability; but there are degrees of testability: some theories are more testable, more exposed to refutation, than others; they take, as it were, greater risks.
6. Confirming evidence should not count *except when it is the result of a genuine test of the theory*; and this means that it can be presented as a serious but unsuccessful attempt to falsify the theory. (I now speak in such cases of "corroborating evidence.")
7. Some genuinely testable theories, when found to be false, are still upheld by their admirers—for example by introducing *ad hoc* some auxiliary assumption, or by re-interpreting the theory *ad hoc* in such a way that it escapes refutation.

    Such a procedure is always possible, but it rescues the theory from refutation only at the price of destroying, or at least lowering, its scientific status.

A good example of what is involved in a valid test as specified by Popper's criteria is provided by what happened in 1818. Augustin Fresnel presented a paper to the French Academy of Sciences as part of a competition. His paper advocated in favor of the wave theory of light that had been introduced by Thomas Young in 1803 at a time when Newton's theory that light consisted of particles had been accepted for over a century. Fresnel argued that when a beam of light encountered an object, then its wave nature would cause it to "bend" around the obstacle, albeit very slightly, similar to the way that sound waves can bend around corners and thus people's voices can be heard even when we cannot see them. The bending of light was forbidden if light consisted of particles because particles traveled in straight lines.

One of the judges of the competition, the eminent statistician and member of the Academy Simeon Poisson, said that the wave theory was manifestly ridiculous because if true, then using symmetry arguments, one should expect to see a bright spot at the very center of the shadow region behind a circular obstacle to a beam of light, something that no one had seen and which seemed to be patently absurd, so absurd that no one had even thought to do such an experiment before. But Dominique Francois Arago was inspired to test Poisson's assertion that the wave theory was obviously false and carried out a very careful experiment and, lo and behold, there was indeed a bright spot at the center of the shadow region. This outcome was so surprising and unexpected that it is now heralded as one of the key experiments that resulted in the rejection of the particle theory of light that had predicted a dark center. By a curious quirk of scientific history, that bright spot was for a while referred to as Poisson's Spot after the person who thought that it would falsify the wave theory, rather than after the two people (Fresnel and Arago) who thought the theory right. But this assignation is not entirely inappropriate because it does take deep insight to think of an experimental test that is so counterintuitive to conventional wisdom, the way that Poisson did.

There is much to like about Popper's formulation of falsification, which is one reason that it has gained widespread acceptance, especially within the scientific community. For starters, it is simple, logical, and easy to understand. While Popper's goal was to create viable demarcation criteria, and his primary motivation for introducing the idea of falsifiability may well have been to foil the attempts of Marxists and psychoanalysts and psychologists from claiming the prestige of science for their theories by only pointing to their "successes," his approach also had the benefit of overcoming the two main problems of the confirmatory approach: that it was impossible to confirm that a proposition had been shown to be true except in a few cases, and that there was no way of specifying how many confirming instances, out of the infinite number possible, were necessary to prove the truth of a theory. Falsifiability claimed that it was possible to unambiguously prove a theory to be false and that a *single* definitive counterexample was sufficient to do so. Using Popper's approach, Arago's experiment, all by itself, could be interpreted as having falsified the particle theory of light.

In addition to seeming to provide a clear and unambiguous test of a theory, falsification also seems to explain much of scientific practice. Scientists do not spend much time in areas of science that have already been carefully examined and found to contain no disagreements with the predictions of theory. This is because it is of little use to repeatedly perform similar experiments and try to build up confirmatory evidence, as a probabilistic approach to proving theories to be true might suggest doing. Such repetitious activity is left to engineers and technicians who seek to apply scientific knowledge to useful practical ends. Instead, scientists focus their energies on those areas that are relatively untested and represent the frontiers of the theory. They seek to push the theory to the limits of its predictive power and use the best available technology to probe these frontiers, in the belief that that is where there are more likely to be breakdowns, the kind of genuine tests that Popper placed such high value on. If the theory does not fail such stringent tests, then we have benefitted by gaining more knowledge about previously unknown applications of it and thus have broadened the range of the theory. But we also benefit (and it is usually more exciting) when the theory unambiguously *fails* a test because such an event signals the need for a newer and better theory, and such improved theories are the main drivers of scientific progress.

Thus the community of scientists is usually working on two different fronts. Some seek to solve the problems and answer the questions that current theories generate and bring existing theory and experiment into agreement. But at the same time others, usually a minority because it is riskier, are on the lookout for areas of research that they feel have the possibility of proving existing theories to be wrong. These two activities are not mutually exclusive since it is often the attempt to get agreement between theory and experiment that throws up instances where such agreement cannot be obtained. These anomalies become the pool from which a falsification candidate emerges. This also explains why, to the mystification of nonscientists, scientists will abandon activity in areas that are perceived to be well established and spend their time on research in esoteric areas that seem to have no practical benefits whatsoever. This does not arise from a self-indulgent desire for intellectual gratification at all costs, a sense of whimsy, or even perverseness. Scientists, like anyone else, would like their work to benefit society and indeed make their living by producing such useful knowledge. But they believe that major long-term benefits come indirectly, chiefly as a consequence of *finding newer and better theories* and proving existing theories to be false is the best way to do so.

Because new frontiers of knowledge are reached less by confirming and expanding old theories and more by the emergence of new theories as a result of old ones being shown to be no longer viable, the failure of a currently accepted theory is not necessarily a cause for sorrow in the scientific community, except perhaps for the original proponents of the now-dethroned theory. Instead, it is very often a cause for excitement. In fact, the more venerable the theory that is suspected of being false, the more exciting it is because such a change is perceived as signifying the need for a dramatic advance in our understanding of nature and often acts as the spur for inventing new theories. As John Carew Eccles said, "I can now rejoice even in the falsification of a cherished theory, because even this is a scientific success" (quoted in Popper 1965, 2). Nothing stimulates the search for better theories more than showing that existing theories are inadequate and possibly wrong. But this focus on highly esoteric areas does leave scientists vulnerable to ridicule by grandstanding anti-science politicians who can point to taxpayer grants given for research that, at least on the surface, seem to provide no public benefit whatsoever.

As an example, recall the intense effort that was put into the search for the Higgs particle. This particle was predicted to exist as far back as the 1960s by the best theory of particle physics we have to date, one that is so widely accepted that it is called the Standard Model, and confirmation of the existence of the Higgs in 2012 was seen as a vindication of the Standard Model. There was no guarantee that it would be found and so the discovery is rightly seen a success. But it also produced mixed feelings. It is nice to know that one is on the right track and there was considerable relief that there was something positive to show for all the money and time and effort invested in the project. Many governments had invested huge sums in building and supporting the Large Hadron Collider in order to find the Higgs and governments tend to be conservative institutions that like to show "success" for their investments. But the greatest success for a scientist lies in creating a new theory so there was also a sense of disappointment when the Higgs was found because finding a confirming instance of an existing theory is not nearly as exciting as finding that an established theory may be wrong. If the Higgs had continued to be elusive, like the aether more than a century earlier, that would have provided greater impetus to researchers searching for new theories to replace the Standard Model.

The drive to find disagreement between the predictions of theory and experiment also explains why scientists will go to extraordinary lengths to create ever more precise measuring instruments, even at great cost. For most everyday purposes, it seems reasonable that once we have reached a useful level of precision in something, there is no reason to go further since the law of diminishing returns makes any additional benefits not cost-effective. But for scientists, the view is quite different. The more precisely they can measure things, the more likely they are to be able to prove something *false*, and so increasing precision is an ongoing goal.

Recall our earlier example of $F = ma$. We said that if each could be measured up to an accuracy of 1%, and if m was measured to be 100 and a was measured to be 10, then the range of values for F would have to lie outside the range 980 to 1020 to enable us to say that the law was false. But if we could increase the level of precision of each measurement from 1% to 0.1%, them m would now lie in the more limited range 99.9 to 100.1 and a would lie in the range 9.99 to 10.01. The range of possible values of ma would now be between 998 and 1002, far narrower than the previously allowed range of 980 to 1020. This narrowing of allowed values for F makes the test of the law more stringent and the likelihood of the theory failing the test becomes much greater, thus also increasing the theory's credibility if it should pass the test. Hence scientists will push the limits of precision as far as human ingenuity, technology, and financial resources will permit.

The idea that we should try to falsify theories rather than trying to prove them to be true has been seized upon by practicing scientists because it seems so clear and unambiguous. It also provides a more focused way of doing science and seems to explain much of scientific practice. It tells scientists that they have to *try to find weaknesses* in existing theories. These weaknesses are initially manifested as anomalies in the data that suggest disagreement with existing theory. Scientists then probe those anomalies more carefully and with greater precision. Very often that closer scrutiny will find fault with the earlier work that indicated a discrepancy, either because of experimental error or because the theory was not applied correctly. The new work then results in the anomaly disappearing and the theory surviving. Such a result usually provides deeper insight into the theory and its applicability and gives us greater confidence in its robustness because it satisfies Popper's criteria of what constitutes a proper test of the theory. But if the anomaly persists and becomes more acute, it may signal a fatal weakness that can lead to the eventual overthrow of the theory.

Another way that falsification guides scientific practice is that it encourages scientists who propose new theories to make predictions that are surprising or, to use Popper's words, those that are "risky," especially predictions that would have never occurred to people working with the older theory. One example is the bright spot of light predicted by Poisson and found by Arago. Another much-cited example is that when Albert Einstein proposed his general theory of relativity in 1915, he predicted that one consequence was that the path of light would be bent by the presence of a gravitational field. According to scientific folklore, it had never occurred to anyone to look for such an effect since existing theory said that light was a wave and hence had no mass and thus should be unaffected by the force of gravity. The reason that no one had observed any deflection of light even accidentally in the course of doing other light experiments was because gravity is such an extraordinarily weak force that its effects are usually tiny and swamped by all the other forces we encounter in everyday life. But it could be magnified to a measurable level by the presence of a massive object like the Sun that creates a huge gravitational field. This idea led to the formulation of an actual test and the reported confirmation of the effect by Arthur Eddington's team in 1919 went a long way toward the acceptance of Einstein's theory. As in the case with Arago a century earlier, no one would have even thought to undertake Eddington's observations before Einstein's theory was proposed, satisfying Popper's criterion of it being a novel and risky prediction. The bending of light waves has since been seen many times and is one of the means by which black holes, massive objects in the universe that are not visible but produce very large gravitational fields, are detected.

As with much of science folklore, this widely accepted dramatic account is not quite accurate, another myth-story that over time has congealed into a scientific fact. Isaac Newton had earlier raised the possibility of light being deflected by gravity because he thought that light consisted of particles and thus could have mass. But the speed of light was not well known in his time nor was its mass, and so Newton left the calculations of the amount of bending as an exercise for others to pursue later. Johann G. von Soldner, not a professional scientist but largely a self-taught man, addressed the problem and published his result in 1801 (Jaki 1978; Soares 2009).

Irony was lurking in the background when Einstein declared in 1907 in a tone of unmistakable originality that gravitation must have an influence on the path of light and gave a formula for its deflection per centimeter. More irony was in store when in 1911 Einstein calculated the bending of light around the sun, because it occurred to him that the effect could be observed during total solar eclipse. Neither Einstein, nor the many readers of his article in *Annalen der Physik*, realized that a hundred and ten years earlier almost exactly the same value, 0."84 versus 0."83 as given by Einstein, had already been calculated by Soldner.

...

But Soldner's article made no ripple in the scientific world. It soon became so forgotten that his biographers, none of them astronomers and cosmologists, could hardly be fascinated by its title and discover its significance. (Jaki 1978)

Einstein's first foray into addressing this problem in 1911 used an earlier version of general relativity known as the Equivalence Principle and it was only later in 1915 that Einstein obtained the correct value using his general theory of relativity. Fortunately for Einstein, his revised calculation was done *before* the Eddington measurements were carried out, as Kevin Brown recounts in his book *Reflections on Relativity*.

The idea of bending light was revived in Einstein's 1911 paper "On the Influence of Gravitation on the Propagation of Light" . . . There were several attempts to measure the deflection of starlight passing close by the Sun during solar eclipses to test Einstein's prediction in the years between 1911 and 1915, but all these attempts were thwarted by cloudy skies, logistical problems, the First World War, etc. Einstein became very exasperated over the repeated failures of the experimentalists to gather any useful data, because he was eager to see his prediction corroborated, which he was certain it would be. Ironically, if any of those early experimental efforts had succeeded in collecting useful data, they would have proven Einstein *wrong*! It wasn't until late in 1915, as he completed the general theory, that Einstein realized his earlier prediction was incorrect, and the angular deflection should actually be *twice* the size he predicted in 1911. Had the World War not intervened, it's likely that Einstein would never have been able to claim the bending of light (at twice the Newtonian value) as a prediction of general relativity. At best he would have been forced to explain, after the fact, why the observed deflection was actually consistent with the completed general theory. Luckily for Einstein, he corrected the light-bending prediction before any expeditions succeeded in making useful observations. (Brown 2017, chapter 6.3)

Another prediction of Einstein's general theory of relativity is that gravitational waves, similar to light and sound waves, should be produced when there is a change in the source of gravity. This too was a risky prediction in the Popperian sense but the problem is that gravitational waves are exceedingly weak, making their detection extremely difficult, and many early attempts to do so failed. But in September 2015, one hundred years after Einstein proposed his theory, scientists finally detected gravitational waves. The result was greeted with relief and celebration as a successful test of a major scientific theory that required overcoming huge technological challenges. Although it took a century to detect the waves, there was no real sense of surprise because, as with the detection of the Higgs particle, confidence in the underlying theory was so great.

What made this detection possible was the building of a pair of extremely large and sensitive L-shaped detectors each about 4 km long in the states of Washington and Louisiana. Scientists then had to wait for some cosmic event on a massive scale to generate a gravitational wave signal powerful enough to be detected. This happened when two black holes with masses 29 and 36 times that of the Sun merged 1.3 billion years ago in the Southern Hemispherical sky to produce a single spinning black hole. In the process, a mass about three times that of the Sun was converted to gravitational energy in a fraction of a second and this was the source of the waves that were detected. The use of two identical detectors separated by about 3,000 km served two purposes: both receiving identical signals would eliminate the possibility of some spurious local effect that could mimic a gravitational wave, and the separate detection locations enabled scientists to locate where in the universe the waves originated (Abbott et al. 2016; LIGO 2016).

At this point a digression is warranted to combat a serious charge made by those who seek, for various motives, to undermine specific scientific theories. They argue that scientists conspire to hide the weaknesses of some of their venerable theories in order to preserve them from being overthrown. This view is promoted by some religious creationists who fear that the theory of evolution by natural selection proposed by Darwin threatens the foundations of their beliefs. It is also promoted by climate change deniers who do not want any controls on their industrial practices that might harm their business interests. Both groups seek to challenge the scientific consensus and argue that scientists act like a cabal with a secret agenda, using their institutional muscle to suppress alternative theories that challenge the scientific consensus on the theory of evolution and climate change models.

This claim is nonsense. First of all, the open way that science is practiced and the large numbers of people involved in arriving at a consensus makes such collusion almost impossible to pull off without being exposed. Furthermore, scientists are often highly competitive and ambitious people. In the scientific community, the greatest prestige does not accrue to those who provide corroborating evidence for existing theories. It is true that if one were able to resolve a major anomaly that was undermining confidence in an existing theory, that would be considered a significant achievement. But the greatest honor goes to those who *overthrow existing theories* and create an entirely new way of under-standing the world. The names of Copernicus, Newton, Darwin, and Einstein will live through the ages because they achieved precisely that kind of scientific revolution. It would be the dream of any scientist to join that pantheon and such people are not going to suppress a revolutionary idea just in order to confirm current theories. If members of the scientific community had even the slightest suspicion that a venerable scientific theory had some kind of fatal flaw, the race would be on to be the first to find and demonstrate it because doing so leads to glory, honors, prestige, and immortality.

It is undoubtedly true that overthrowing a well-entrenched scientific theory is not easy (and there are excellent reasons as to why this should be so as I will discuss in chapter 12) but it is the strength of science that if one has credible evidence and arguments in support of a new theory, one will be listened to by at least some of one's peers. Later on we will explore further how the community of adherents to a new theory grows and old theories get overthrown.

While the idea that you cannot affirmatively prove theories to be true may be discouraging, the falsification idea of Popper that you can unambiguously prove theories to be false is appealing because it seems to provide a way of *backing* into true theories. If we can systematically eliminate false theories, then we are not only preventing people from wasting time and effort and money pursuing blind alleys, surely we are getting closer to finding true theories, provided one understands the meaning of "true" to mean "the one re-maining theory that has not yet been shown to be false." This reasoning is captured by the dictum of Arthur Conan Doyle's famous fictional detective Sherlock Holmes who said in *The Sign of Four* that "When you have elimi-nated the impossible, whatever remains, however improbable, must be the truth."

It is not surprising that for all these reasons, the idea of falsification has taken deep root within the scientific community and even the general public as being almost self-evidently true, not only as the way that scientific theories evolve but also as a satisfactory demarcation criterion, the marker that distinguishes sci-ence from nonscience. As a result, Popper is the philosopher of science whose work is best known amongst them and quoted extensively and approvingly. So, for example, astrophysicist Mario Livio can write the following passage confi-dent that fellow scientists would accept it as a truism:

> Ever since the seminal work of philosopher of science Karl Popper, for a sci-entific theory to be worthy of its name, it has to be falsifiable by experiments or observations. This requirement has become the foundation of the "scientific method." (Livio 2013, 262)

Unfortunately, far from being the foundation of the scientific method, falsification turns out to be untenable. The problem is that falsifying a theory is not as easy as it sounds and in fact turns out to be impossible *even in principle* because it depends upon a clear distinction between theory on the one hand and experimental or observational facts on the other, with the latter being completely objective and incontrovertible. If that were the case, then one can indeed compare the predictions of the theory with the facts and if they disagree, then the theory is falsified. This seemed to be what happened with the particle theory of light as a result of the detection of Poisson's Spot. But such a clean separation is not possible because to obtain facts that are outside the range of our senses requires the involvement of *other* theories as well, since most scientific facts are obtained inferentially, and this complicates the picture.

Recall that any measurement outside the range of our senses is essentially a reading on an instrument. To convert that reading into something meaningful requires additional theories. For example, suppose we seek to test Ohm's Law for a wire, a common experiment that is done when students are introduced to electricity. This law asserts that for certain types of materials, the amount of electrical current flowing through a wire made of that material rises proportionately with the voltage applied across its ends. Testing this law requires us to measure the values of the current through the wire and of the voltage across the ends of the wire. To carry this out we can use ammeters and voltmeters that have pointers or digital displays that give us numbers for each quantity. These numbers are the raw material of our data. However, to interpret the voltmeter numbers as voltages and ammeter numbers as currents requires us to have a theory of how these meters work and this in turn involves the use, at a minimum, of theories of electromagnetism and mechanics. So the measured values of current and voltages are themselves not theory-independent objective facts directly obtained by instruments but can be considered as *theoretical* constructs far removed from *sense-data*, which are those things that we can see, touch, feel, smell, and hear.

The fact that theories are always involved in the creation of experimental facts can be used by those opposed to some scientific conclusion to deny the validity of scientific results that they find uncongenial. For example, it used to be thought in medieval times that celestial objects, being part of the heavenly firmament, must be perfect and consist of perfectly smooth spheres. But when Galileo turned his telescope toward the Moon he found features that no one had seen before, leading him to state: "By oft-repeated observations of them we have been led to the conclusion that we certainly see the surface of the Moon to be not smooth, even, and perfectly spherical, as the great crowd of philosophers have believed about this and other heavenly bodies, but, on the contrary, to be uneven, rough, and crowded with depressions and bulges. And it is like the face of the earth itself, which is marked here and there with chains of mountains and depths of valleys"

(Frova and Marenzana 2006, 162). He also found that the planet Jupiter had its own orbiting satellites, which challenged the conventional geocentric view that every celestial body had to orbit the Earth.

This news shocked people of that time and some tried to retain the idea of celestial perfection by arguing that it was the *telescope* that had introduced the rough effects that were not visible to the naked eye and thus those observations were not a valid representation of reality. This view was made more credible by the fact that in those days the lenses were crude and introduced considerably more distortions than modern ones. The story is told of the eminent and highly respected Aristotelian philosopher Cesare Cremonini refusing to even look through the telescope, giving as his reason that it was unnecessary because Aristotle had proved that the Moon had to be a perfect sphere. While this story is told as an example of blind allegiance to authority, the truth, as is so often the case, may be more complicated, and shows the difficulty of ascertaining the reliability of historical events. One alternative story is that Cremonini was a friend of Galileo's and was aware of the danger that Galileo was in from the Inquisition because of his heretical views. Not looking through the telescope may have been his way of avoiding having to take a stand that would have required him to either contradict his friend or incriminate himself for the same offenses as Galileo. Another alternative story is that he did indeed look through the telescope but that because of the crudity of the lenses, he did not see clearly what Galileo claimed and that instead the effort gave him a headache.

We now have much better lenses and well-established theories of optics that show that what the telescope basically does is magnify sizes without introducing major new features, apart from some minor and incidental distortions. But this means that what one "sees" through a telescope depends not just on sensory data but also on the correctness of the optical theories underlying the working of the telescope, and the question can be shifted to that of deciding how we know that *those* theories are true. Determining the correctness of the optical theories of the telescope involves using other theories, and so on. Moving away from sensory data can lead to such endless regressions. But even if we had data that we directly received via our senses, say that of sight, that would not eliminate all theory dependence because it can be argued that even our sensory experiences involve theories. We would still need a theory of light and vision to relate the signals our eyes receive with that emitted by the object and the message that our brain creates. After all, when I say I "see" something, what I mean is that some external stimulus enters through the cornea and lens of my eye and falls upon my retina that then converts it into a nerve signal that is transmitted to the visual cortex of my brain that then interprets that signal and manifests itself as an impression in my consciousness. That is quite a complicated process involving many theories.

Popper himself later realized this.

> Sense-data, untheoretical items of information, simply do not exist. For we *always* operate with theories, some of which are even incorporated in our physiology. And a sense-organ is akin to a theory: according to evolutionist views a sense-organ is developed in an attempt to adjust ourselves to a real external world, to help us to find our way through the world. A scientific theory is an organ we develop outside our skin, while an organ is a theory we develop inside our skin. This is one of the many reasons why the idea of completely untheoretical, and hence incorrigible, sense-data is mistaken. We can never free observation from the theoretical elements of interpretation. We always interpret; that is we theorize, on a conscious, on an unconscious, and on a physiological level . . . We *cannot* justify our knowledge of the external world; *all* our knowledge, even our observational knowledge, is theoretical, corrigible, and fallible. (Popper 1968b, 163–64, emphasis in original)

It should be emphasized that the fact that pure sense-data do not exist does not contradict the fact of the reliability and repeatability of science's empirical basis. All it means is that the basis of falsification, that one can use supposedly objective experimental or observational data to eliminate false theories, turns out to be not so clear-cut because the unavoidable "contamination" of the data by theory prevents such a clear demarcation. What we really have is a conflict between the theory ostensibly being tested and the theories that are inextricably entwined with the experimental observations, and so the issue becomes one of deciding which theories we wish to accept and which we need to modify or even reject. *How choices are made between competing theories turns out to be the fundamental issue in science epistemology.*

But there is yet another serious objection to the idea of falsification and that is that while it does explain much of how scientists go about their work, it does not describe how scientific theories have actually evolved. If applied strictly, it would be disastrous for science because no theory ever agrees with *all* the data it confronts. There are *always* anomalies and it is the investigation of these discrepancies that occupy much of scientific research. In general, new theories usually agree with just a few observations and it takes much hard work by dedicated people to slowly get greater agreement with more data, though some disagreements always remain. If the falsification rule were strictly applied, *every theory would be considered immediately falsified and required to be thrown out*, even those theories that we now hold up as representing the best of science. To dislodge the firm hold that falsificationism has on the minds of people about how science works, I shall provide several examples of how some of the most famous theories have existed alongside major anomalies sometimes for centuries without being considered falsified, thus demonstrating that falsification does not explain the history of scientific evolution.

For example, Newton's theories of motion and gravitation did not predict exactly the motion of the planets nor explain the stability of the solar system. The reason was that while it was fairly straightforward to apply his theories to the motion of a single planet under the gravitational field of the Sun, this was not the whole story. That simple model gave approximately correct results because by virtue of its great mass, the Sun is by far the source of the largest gravitational force in our planetary system. But the planets create gravitational forces as well and the effects of those forces on each other vary as the planets move around in their orbits. Incorporating those effects involves extremely difficult calculations, well beyond the capabilities of Newton or the scientists who came after him. So the question Newton had to address was whether these extra forces somehow averaged out to produce no net effect over the long term or whether their effects accumulated over time and made the system unstable, resulting in the planets beginning to wobble in their orbits and eventually causing the system to collapse. He came to the conclusion that the solar system was unstable but that God periodically intervened to press some kind of cosmic reset button when the orbits threatened to get too eccentric.

Of course, such a supernatural solution to the problem of planetary stability was hardly satisfactory and Newton's suggestion came in for some ridicule. As Scott Tremaine writes:

> Newton's comment on this problem is worth quoting: "the Planets move one and the same way in Orbs concentrick, some inconsiderable Irregularities excepted, which may have arisen from the mutual Actions of Comets and Planets upon one another, and which will be apt to increase, till this System wants a Reformation." Evidently Newton believed that the solar system was unstable, and that occasional divine intervention was required to restore the well-spaced, nearly circular planetary orbits that we observe today. According to the historian Michael Hoskin, in Newton's world view "God demonstrated his continuing concern for his clockwork universe by entering into what we might describe as a permanent servicing contract" for the solar system.
>
> Other mathematicians have also been seduced into philosophical speculation by the problem of the stability of the solar system. Quoting Hoskin again, Newton's contemporary and rival Gottfried Leibniz "sneer[ed] at Newton's conception, as being that of a God so incompetent as to be reduced to miracles in order to rescue his machinery from collapse." A century later, the mathematician Pierre Simon Laplace was inspired by the success of celestial mechanics to make the famous comment that now encapsulates the concept of causal or Laplacian determinism: "An intelligence knowing all the forces acting in nature at a given instant, as well as the momentary positions of all things in the universe, would be able to comprehend in one single formula the motions of

the largest bodies as well as the lightest atoms in the world, provided that its intellect were sufficiently powerful to subject all data to analysis; to it nothing would be uncertain, the future as well as the past would be present to its eyes. The perfection that the human mind has been able to give to astronomy affords but a feeble outline of such an intelligence." (Tremaine 2011)

It is possible that part of the reason for Newton's suggestion that God had to periodically intervene was due to a fear arising at that time that a purely mechanical universe that functioned perfectly would make God superfluous once it had been set it in motion (Butterfield 1957, 125). So such a "permanent servicing contract" had the benefit of making God a necessary explanatory entity.

Some of the most eminent scientists and mathematicians devoted great efforts over the next three *centuries* to solving the immensely challenging problem of the stability of the solar system, initially analytically and more recently using computer models. The current conclusion seems to be that the system is stable at least on the order of billions of years, but we do not yet have a definitive answer. Yet the fact that Newton's theory was not considered falsified because it could not explain the stability of the solar system shows that it is not necessary for a theory to answer every single question, even an important one. The scientific community is willing to live for a long time, centuries even, with a theory that is highly useful and productive even if it can be argued that it has been technically falsified in a few instances.

As another example, recall the situation prior to the discovery of the planet Neptune. It had been known for some time that the motion of some known planets such as Uranus were not quite consistent with Newton's laws which formed the basis of mechanics and gravity during that period. If what Imre Lakatos labels as Popper's "naive falsificationist" model of scientific progress were actually in operation, it would have been argued that this anomalous behavior had effectively disproved Newton's laws and thus they should be scrapped (Lakatos 1970). But that is not what happened. Defenders of Newton's laws (which consisted of almost everyone in the scientific community because of its immense power and success) put down the failure as being due to possible errors in observations or in the calculations or other unknown factors. One can always argue that there are some hitherto unknown or overlooked factors causing the problem. Thus the discrepancy, though persistent, was classified as merely an anomaly, a problem to be solved, and not as a falsifying event. The anomaly was investigated by a few scientists as their chosen area of research while the majority continued to confidently use Newton's laws in their separate areas of research activity.

This situation continued for many years until the discovery of Neptune in 1846 explained the anomalous behavior. In fact, this discovery was itself strongly aided by the belief scientists had that Newton's laws were true because that enabled them to hypothesize that perhaps an unknown planet may be perturbing the motion of the known planets. Aided by this belief, the astronomer Urbain Jean Joseph Le Verrier was able to use Newton's laws to predict exactly where a hitherto unobserved planet would have to be in order to produce the anomalous behavior of Uranus. Ironically, the same Arago who had been inspired to look for and detect Poisson's bright spot was now serving as the director of the Paris Observatory and in that role declined to give Le Verrier any telescope time to search for his planet, seeing it as a wild goose chase (Chown 2018). This forced the latter to appeal to colleagues elsewhere to search on his behalf. The challenge was taken up by Johann Galle at the Berlin Observatory who was able to find Neptune in less than an hour of searching (Sheehan 2003). This discovery converted what could have been considered a falsifying event into a tremendous triumph for Newtonian mechanics. It is the existence of such a strong belief in the correctness of a given theory that gives researchers the confidence that if they work hard at the calculations and observations, they will find the solution to an anomaly, and prevents them from assuming too quickly that a cherished theory is falsified.

Spurred by the success of his prediction of Neptune, it was natural for Le Verrier and others to propose a similar solution to another anomaly, and that was the behavior of the planet Mercury. Its elliptical orbit around the Sun changed slightly each year and the annual shift in its point of closest approach (known as the perihelion) could be calculated using Newton's laws. But observations and calculations done at that time showed a clear and persistent discrepancy of 38 arcseconds per century. To explain this, in 1859 Le Verrier proposed that a new and as yet unseen planet (that was named Vulcan) lay inside the orbit of Mercury. Despite claims by some observers to have seen it, the observations were inconclusive so that by 1878, after another set of careful observations during an eclipse failed to show it, only a few die-hard believers were still looking for it. But its existence could not be definitively ruled out since the discrepancy persisted, and indeed improved calculations and observations had increased the amount of the discrepancy to 43 arcseconds per century. It was only after Einstein published results in 1915 showing that the motion of the perihelion agreed with calculations using his general theory of relativity that Vulcan's existence could be said to have been "disproved," because it had become unnecessary as an explanatory concept (Levenson 2015).

Interestingly, this same pattern is being repeated in our own time. Researchers have observed anomalies in asteroids and other bodies of varying sizes like Pluto that occupy the region beyond Neptune called the Kuiper belt. Again, rather than asserting that this anomalous behavior is due to our current theories of planetary motion being wrong, they have postulated the existence of a massive new planet that is about 1,200 times the mass of the Earth and has an extremely eccentric elliptical orbit that passes through the Kuiper belt and at its greatest distance is between 10,000 and 20,000 times the radius of the Earth's orbit, thus explaining why we have not observed it as yet. But that is not the only explanation for the anomalous behavior and the various alternatives will compete to see which can gather more supporting evidence until one of them becomes the consensus view. Of course, direct observation of the postulated new planet (where "direct" means a telescopic sighting) would definitely swing the argument in its favor.

Another example is from our earlier discussion on the age of the Earth. During the fifty-year period from 1850 to 1900, neither the geology community nor the physics community felt that their own theories that arrived at results that were contradicted by the other had been falsified, although a strict application of the falsification rule would have required this. In science, there is no supra-arbiter to say which of two competing groups needs to abandon their existing theories and find new ones that produce results in conformity with the other group's work. Both branches of science proceeded to regard this discrepancy as an anomaly (although a highly visible and embarrassing one) that would be resolved eventually. Similarly, during the period when astrophysicists were arriving at an age of the universe that was less than the age of the Earth as determined by geologists, neither branch of science felt that their fundamental paradigms had been falsified but continued using them, in the belief that this was an interesting puzzle to be addressed and would eventually succumb to their efforts and be resolved. And they were right.

For yet another example, consider the orbit of the Moon around the Earth. The closest point of this almost circular orbit is known as the *perigee* and this point is not fixed but shifts slightly each year, just like in the case of Mercury around the Sun. Newton's theory of motion failed to agree with the observed motion of the Moon's perigee, getting only half the needed value (Cook 2000). This discrepancy persisted for nearly sixty years. Since this was a well defined and narrowly tailored question, the discrepancy was more puzzling than most and there were suggestions that Newton's law of gravity had been falsified and calls for modifications to its inverse square law. But these calls were resisted and this "faith" (to use that problematic word again) in the correctness of the theory was rewarded when it was shown that it was the incorrect application of mathematics to the problem that had led to the discrepancy (Kuhn 1970, 81). Interestingly, as we will see in chapter 17, calls for modifying Newtonian gravity have surfaced once again in the context of dark matter theories.

Such examples of potentially falsifying results not being treated as such can be multiplied many times. When measurements of solar neutrinos reaching the Earth from the Sun showed only a third of the number predicted by theories of energy production in stars, those stellar energy models were not considered falsified and summarily abandoned. Instead the anomaly was investigated as an interesting research problem. And eventually scientists were able to show that the difference was due to neutrinos on their journey from the Sun to the Earth oscillating from one form (the type that the detectors were designed to register) into another form that was not being detected. Similarly, a new particle with a mass of six times that of the Higgs boson was supposedly found in the Large Hadron Collider in late 2015. Since that did not fit into the Standard Model, it thus caused some excitement because it could have had implications for many areas, such as the search for dark matter, and raised the possibility of the need for a new theory such as quantum gravity. But it was later found to be the product of chance (Scharping 2016).

In general, most scientists treated all these discrepancies not as falsifying events but as anomalies that further research would resolve, and eventually they were. These and many similar examples have led philosophers of science to argue that *all theories* at *all times* have disagreements with experiments that *could* be classified as falsifying events. When a new theory is first invented, it is usually in response to a specific and fairly narrow need that existing theories have not been able to meet. When first created, a theory typically explains only a few experimental results, is untested in most areas, and does not agree with some other results. It takes considerable time and effort on the part of many scientists to flesh out the new theory in a widening range of applications. But all the anomalies are never completely eliminated.

If the Popper criterion of falsifiability were strictly followed, then Newton's theory, and indeed all theories, would be rejected almost as soon as they were proposed. Yet as the many examples quoted here show, it is a fact that the scientific community has held on to certain theories for considerable lengths of time despite the existence of even serious discrepancies with experiments and observations. During each such period of stability, the currently dominant theory continues to have the allegiance of almost all the community within the specialty relevant to the theory. This kind of splitting of tasks, where a few scientists investigate specific anomalies while the majority assume that the theory is correct and continue to use it in other applications in the belief that the anomaly will eventually disappear, is pretty much standard practice in science and is one reason for its success. The appearance of an anomaly does not bring everything to a grinding halt until it is resolved. While anomalies, if seen as serious and persistent, may stimulate the search by some scientists for new theories, by themselves they are not considered as falsifying existing theories.

This "faith" in the rightness of current theories is by no means a sign of irrational dogmatism as critics sometimes assert. Instead it forms the sound foundation for productive scientific practice. It ensures that scientists are not easily swayed into rejecting useful theories by this or that supposedly falsifying experiment that happens to come along. As Kurt Lewin, one of the founders of experiential learning, said, "There is nothing so practical as a good theory" (Kolb 1984, 4), and since good theories are hard to come by, one should not cavalierly discard them. Doing so would greatly inhibit the progress of science. The strong allegiance of the scientific community to existing scientific theories is also paradoxically what enables them to see more clearly when those theories are inadequate, and lays the foundation for the periodic revolutions in which new theories actually do overthrow the old. The problem we face is how to explain such changes in the absence of any objective criteria for verifying or falsifying theories.

Most if not all scientists understand we cannot prove our theories to be true. Many are not as aware that trying to assign relative probabilities to competing theories is also unworkable in practice. But the belief that scientific theories can be proven false is still widely held among scientists and the general public, and convincing them otherwise is not easy because the scientific literature is replete with allegedly falsifying experiments that led to revolutions. This is why I felt obliged to provide so many examples that go against it. The fact is that the idea of falsification, however appealing it may be because of its simplicity, is simply not viable as a model of scientific progress and must be abandoned. To understand how scientific revolutions come about without falsification and yet are so strongly believed to be due to it is discussed in the next section and in chapters 18 and 20.

(c) Scientific revolutions are the result of *three*-cornered struggles

One of the problems with the idea that scientific theories are rejected because of falsification is that if that were true, one would have a period after the old theory has been falsified and before a new one emerges to fill the vacuum during which there would be no scientific theory to work with. In reality, that never happens. There is never a period when scientists are adrift, twiddling their thumbs, waiting for a new theory to come along so that they can get back to work. The reason for the absence of a theoretical vacuum is because scientific revolutions are never caused by a conflict between theory and experiment. Instead, they are always the outcome of *three-cornered* struggles in which two theories, the old and the new, compete for the allegiance of the scientific community, and experimental data forms the third element of the relationship.

When an existing scientific theory is showing signs of stress in that some major anomaly is defying efforts to bring theory into agreement with experiment, researchers become increasingly dissatisfied with it and start loosening its constraints and exploring other avenues more freely. But they are not doing so randomly. They are guided by at least a tentative alternative hypothesis that suggests specific experiments as tests. Scientists do not behave like random fact-gatherers, unconstrained by any expectations of what they hope to find. All expectations are based on some theory, however tentative, despite folklore to the contrary. As Thomas Kuhn explains using the examples of Boyle's and Coulomb's laws:

> Perhaps it is not apparent that a paradigm is prerequisite to the discovery of laws like [Boyle's, Coulomb's, and Joule's laws]. We often hear that they are found by examining measurements undertaken for their own sake and without theoretical commitment. But history offers no support for so excessively Baconian a method. Boyle's experiments were not conceivable (and if conceived would have received another interpretation or none at all) until air was recognized as an elastic fluid to which all the elaborate concepts of hydrostatics could be applied. Coulomb's success depended upon his constructing special apparatus to measure the force between point charges. (Those who had previously measured electrical forces using ordinary pan balances, etc., had found no consistent or simple regularity at all.) But that design, in turn, depended upon the previous recognition that every particle of electric fluid acts upon every other at a distance. It was for the force between such particles—the only force which might safely be assumed a simple function of distance—that Coulomb was looking. Joule's experiments could also be used to illustrate how quantitative laws merge through paradigm articulation. In fact, so general and close is the relation between qualitative paradigm and quantitative law that, since Galileo, such laws have often been correctly guessed with the aid of a paradigm years before apparatus could be designed for their experimental determination. (Kuhn 1970, 28–29)

What of the stories of serendipitous discoveries that permeate the folklore of science, of scientists finding something when they were supposedly not looking for it? Two popular examples are the discovery of X-rays and the cosmic microwave background radiation. But these too occurred within the context of theoretical expectations. They both involve cases where the experimenters had such clear expectations of what results they should obtain that they were able to recognize that something had gone seriously awry and began to investigate possible causes. In the former, Wilhelm Roentgen was working with cathode rays when he noticed that a barium-platino-cyanide screen at some distance from his shielded apparatus started glowing green while he was running his experiment (Kuhn 1970, 57). In the case of the cosmic microwave background radiation, Arno Penzias and Robert Wilson were designing a high quality radio telescope. Their clear expectations of the radiation patterns they should detect when subjecting it to tests enabled them to realize that the puzzling signals they were getting and which they could not eliminate could not be random noise but signified something deeper.

Thomas Kuhn has been highly influential in elucidating the nature of the scientific process and the conditions under which scientific revolutions occur. In doing so, he introduced into the field the idea of a *scientific paradigm*. Over time (and even within the book *The Structure of Scientific Revolutions* where he introduced the concept), the term paradigm has been used with many shades of meaning and interpretation. At the most basic level, a paradigm represents a template for a pattern that can be applied in many different contexts. I will use the word more broadly, in the sense of a *scientific framework* that surrounds a scientific theory or a closely-knit constellation of theories. This framework consists of the rules of operation of the theory or theories, the kinds of problems that are meaningful for scientists to work on, the standards by which solutions are judged for correctness, the people who are seen as reliable judges of whether a solution to a problem or anomaly meets the standards set for resolution, and so on. Broadly speaking, when scientists work within a paradigm, they have confidence that what they do is worthwhile because it is within the bounds set by the community of scientists who share that paradigm.

A good analogy to a scientific paradigm is a constitutional democracy. At its heart lies a written constitution that is analogous to the basic scientific theory. But such a document only enunciates broad general principles and, standing alone, is not sufficient to serve as a guide for how a society should operate. To serve a useful function, the constitution needs to be elaborated upon and fleshed out so that it can be used in concrete instances. Providing that wider structure becomes the role of the legislature and the various other branches of the government. The courts are given the role of arbiters to resolve conflicts of interpretation, and the executive branch can then act on the resulting interpretations of the constitution. Of course, these roles and divisions are not as clearly demarcated as they seem on the surface. In practice there is a lot of overlap due to flexibility and elasticity in the system. This is necessary because a system that is too rigid is more likely to break under stress, whereas a flexible system can adapt to changing times and situations and requires extreme conditions to become unworkable, such as when developments occur that simply cannot be accommodated within the existing system and there are no agreed-upon supra-arbiters to determine the outcome. When that happens, one gets a political revolution, as happened when the colonies of the European powers sought independence around the world. As a specific example, when American colonists decided that rule by England was no longer able to accommodate their own needs and aspirations, there was no body to appeal to that could decide if their right to self-government was justifiable and, if so, how such a transition would occur. The outcome was eventually determined using force.

Kuhn's insight was that scientific paradigms have close parallels to the structures that surround and support a constitutional system, and scientific revolutions have close parallels to political revolutions. Scientific revolutions occur when two incompatible paradigms come into conflict and there is no way within the existing framework to adjudicate which one is correct and should guide future science. A new paradigm often deals with problems and methods that lie outside the scope of the old paradigm, limiting the ability of the old to judge the new paradigm's claim to validity and vice versa, thus making direct comparison impossible. The two paradigms are thus said to be *incommensurable*. A scientific revolution occurs when a new paradigm is able to shift the allegiance of the majority of the scientific community to it, and evidence and arguments play a major role in the process. Of course, in the case of science and unlike with political revolutions, the final outcome of such conflicts is not determined by physical force or violence. But scientific revolutions do share with political ones the feature that the outcomes are not entirely determined by the use of objective criteria.

In the case of a successful political revolution, some of the structures and rules and personnel that served the specific needs of the old regime may no longer be necessary or desirable and one usually requires new ones to replace the old. But people are naturally reluctant to abandon what was so familiar and useful for so long. So it is with scientific revolutions. Once a paradigm has been found to be fruitful and sufficiently correct in its predictions over a period of time, scientists are loath to abandon it because to do so would be to also abandon many of its hard-won successes, not to mention many of the skills painfully acquired in order to address the problems that the paradigm threw up.

This belief in the rightness of an existing paradigm and the strong desire to retain it should not be viewed as irrational dogmatism. As Kuhn said, it is what enables scientists to distinguish when a theory is *truly* in crisis.

> Novelty ordinarily emerges only for the man who, knowing *with precision* what he should expect, is able to recognize that something has gone wrong. Anomaly appears only against the background provided by the paradigm. The more precise and far-reaching that paradigm is, the more sensitive an indicator it provides of anomaly and hence of an occasion for paradigm change. In the normal mode of discovery, even resistance to change has a use . . . By ensuring that the paradigm will not be too easily surrendered, resistance guarantees that scientists will not be lightly distracted and that the anomalies that lead to paradigm change will penetrate existing knowledge to the core. (Kuhn 1970, 65, emphasis in original)

The reason that scientific revolutions, like political revolutions, are rare is that they are often very costly, resulting in losses as well as gains. It is the nature of scientific revolutions that new paradigms do not preserve all the knowledge generated by the old. The status of solutions to problems that had been considered solved under the old paradigm now become uncertain or even unsolved under the new. For example, as Kuhn points out, "Lavoisier's chemical theory inhibited chemists from asking why the metals were so much alike, a question that phlogistic chemistry had both asked and answered. The transition to Lavoisier's paradigm had, like the transition to Newton's, meant a loss not only of a permissible question but of an achieved solution" (Kuhn 1970, 148). Similarly, all the explanations based on affinity chemistry went away with the arrival of atomic chemistry (Laudan 1977, 113). As another example, the abandonment of the aether meant that all the work that had gone into measuring its properties such as its density, elasticity, and refractive index was no longer of any value. As we will see in chapter 20, the replacement of the Ptolemaic model of the solar system by the Copernican one resulted in many questions that were thought to have been answered (such as the directions of "up" and "down" and why an object thrown vertically upward returns to the point of origin) became open questions again. Those who played key roles in generating the successes of the old paradigm will naturally be reluctant to adopt the new one and may even actively resist it, since much of their prior hard work and successes will seem to have been for naught.

How scientific revolutions eventually come about is a complicated process that will be discussed in more detail in chapter 18, but a necessary condition is the presence of a *competing theory* that ultimately determines whether a discrepant event is treated as merely an anomaly that should be solvable under the old paradigm or as a falsifying event that requires the new one. As long as no promising competing theory exists, such discrepant events are treated as anomalies. When a viable contending theory appears and manages over time to command the allegiance of most members of the scientific community to work within it, then that same event may be elevated to the status of a falsifying event as part of the effort to persuade scientists to switch their allegiance.

One important consequence of this result is that we have to relinquish the notion that truth and falsity are arrived at purely objectively and, once established, are eternally true, as had been the hope at one time. Scientific paradigms are never *proven* to be true or false but instead are *deemed* to be true or false as a result of an informal yet powerful consensus judgment by the community of scholars in the field. As sociologist of science Barry Barnes describes it, "In agreeing upon a paradigm scientists do not accept a finished product: rather, they agree to accept a basis for future work, and to treat as illusory or eliminable all its apparent inadequacies and defects" (Barnes 1982, 46) and "In science . . . there is no basis for validation superior to the collective contingent judgment of the paradigm-sharing community itself" (Barnes 1982, 50–51).

The repeated overthrow over time of accepted scientific theories in favor of new ones is thus not a source of concern but merely reflects the fact that scientific truths are at best *provisional*, the best we have at the moment. What we believe to be true now is always subject to revision and may well turn out to be judged false in the future. Even if by some chance we happened to stumble upon the one true theory or theories (assuming that such things exist at all), there is no way to conclude for sure that we have done so. It may pass test after test and last for a long time but that alone is insufficient to think of it as definitely true, since other theories have also lasted for a long time in the past before being overthrown. The problem of induction inevitably raises its head and leaves open the possibility that the very next test we carry out may throw up an anomaly that over time turns into a serious crisis that results in the overthrow of a long-standing theory. We cannot rule out the possibility that this might happen and this precludes any claims of absolute certainty.

We have to learn to live with the fact that scientific theories, however broad they are in scope, however widespread and powerful they are in their applications, however successful they are in producing results, and however durable they have been, are always only provisionally true.

(d) Scientific theories are always underdetermined by data

In the context of falsification, I quoted earlier Arthur Conan Doyle's famous fictional detective Sherlock Holmes who said in *The Sign of Four*, "When you have eliminated the impossible, whatever remains, however improbable, must be the truth." We have seen that in the scientific context, "eliminating the impossible" by the method of falsification cannot be done. Proving a theory false just by comparing its predictions with data turns out to be an illusion, thus ruling out a simplistic application of Holmes's method. But it is possible to make a *reasoned judgment*, based on a preponderance of evidence, that a theory is false. That would seem to provide the hope that one could, by repeatedly using such reasoned judgments, eliminate all false theories and thus be left with just one unfalsified theory that can be judged to be true. While it may not have been objectively shown to be true, the fact that it is the sole theory left standing, and provided that it has considerable empirical evidence in support of it, surely adds weight to the idea that it may well be the one true theory.

The problem is that however many theories you judge to be false, you cannot zero in on a true theory. The truth of Holmes's statement is crucially dependent upon there being just a *finite* number of possible theories, so that one can arrive at the true one by systematically eliminating all the false ones. This turns out to not be possible because *theories are always underdetermined by the data*. Pierre Duhem pointed out that science differs from mathematics in that one *never* exhausts all the possible scientific theories that are possible in any given context. No given set of data *uniquely* determines a theory that explains that data. However large and varied the set of data that one seeks to explain, there will always be an *infinite* number of possible theories that serve the purpose.

> Unlike the reduction to absurdity employed by geometers, experimental contradiction does not have the power to transform a physical hypothesis into an indisputable truth; in order to confer this power on it, it would be necessary to enumerate completely the various hypotheses which may cover a determinate group of phenomena; but the physicist is never sure that he has exhausted all the imaginable assumptions. The truth of a physical theory is not decided by heads or tails. (Duhem 1906, 190)

Hence even if the process of falsification were able to weed out wrong theories, the idea that by repeatedly using this process of elimination we will end up with just one theory left standing that will be the true one is illusory. But the desire to think that our current theories are uniquely forced upon us by the weight of evidence is strong, because it gives those theories a heft that they might otherwise lack. For example, when he proposed his laws of motion and gravitation, Isaac Newton made the claim that his laws were forced on him by the data and thus free of any theoretical speculation.

> Newton himself thought that he proved his laws from facts. He was proud of not uttering mere hypotheses: he only published theories proven from facts. In particular, he claimed that he deduced his laws from the "phenomena" provided by Kepler. But his boast was nonsense, since according to Kepler, planets move in ellipses, but according to Newton's theory, planets would move in ellipses only if the planets did not disturb each other in their motion. But they do. This is why Newton had to devise a perturbation theory from which it follows that no planet moves in an ellipse. (Lakatos 1973, 2)

People tend to believe that data can uniquely constrain scientific theories and can be used in situations where there are two or more competing theories that purport to explain a set of data. It is true that if one takes each of those theories and makes predictions for some future experiment and then carry out that experiment, the result may result in the elimination of some or even all but one of the competing theories. But there will also emerge a new and different set of alternative theories that can explain the enlarged data set. While one can invoke other criteria to reject all but one theory, these will be based on subjective criteria such as aesthetics and there are no compelling reasons that compel us to use them. Newton's claim that his theory was purely data-driven was false. Theories are always underdetermined by data, however large the data set may be, because we can always find an *infinite* number of alternative theories that explain the same set of data. (In the Supplementary Materials B section at the end of this book, I give examples from mathematics as analogies that may help in understanding why Newton had to be wrong, since it is in that field where the concept of *proving* something to be true is strongest.)

Imre Lakatos argues that it is possible that Newton may not have really believed that his theories were forced upon him by the data, that given the climate in which Newton operated, his statements may not have represented his genuine feelings but were merely a clever argumentative move on his part to forestall possible objections.

> Scientists want to make their theories respectable, deserving of the title "science," that is, genuine knowledge. Now the most relevant knowledge in the seventeenth century, when science was born, concerned God, the Devil, Heaven and Hell. If one got one's conjectures about divinity wrong, the consequence of one's mistake was eternal damnation. Theological knowledge cannot be fallible: it must be beyond doubt. Now the Enlightenment thought that we were fallible and ignorant about matters theological. There is no scientific theology and, therefore, no theological knowledge. Knowledge can only be about Nature, but this new type of knowledge had to be judged by the standards they took over straight from theology: it had to be proven beyond doubt. Science had to achieve the very certainty which had escaped theology. A scientist, worthy of the name, was not allowed to guess: he had to prove each sentence he uttered from facts. This was the criterion of scientific honesty. Theories unproven from facts were regarded as sinful pseudoscience, heresy in the scientific community. (Lakatos 1973, 2)

But Newton historian Rob Iliffe argues that Newton actually did believe that his conclusions were utterly factual. Unlike Einstein who said, "Imagination is more important than knowledge. Knowledge is limited. Imagination encircles the world" (Calaprice 2005, 9), Newton felt that imagination was a trap that misled the unwary. He condemned the use of hypotheses because he felt that they would lead to "interminable and barren disputes" (Iliffe 2017, 14) and took umbrage at any suggestion that his own work could be considered to contain hypotheses that were merely probable and not certain.

Newton was hostile to the idea that scientific knowledge could only at best be probable, and he believed that it was capable of the same level of (absolute) certainty as was found in the mathematical sciences. (Iliffe 2017, 316)

...

Newton's account of the appropriate scientific roles, first of disciplined and rational enquiry, and second, of hypotheses, was an expression of the way that he understood that his scholarly self should best be ordered ... [He] argued that the hard work that disciplined the understanding could control the temptations and seductions of the imagination. The triumph of the imagination drew the godly man away from righteousness, just as the introduction of hypotheses into natural philosophy corrupted the pursuit of truth. (Iliffe 2017, 351)

Newton's theory of gravitation also seemed, by the standards of that time, to be bordering on the occult since he was arguing that space consisted of a vacuum and that inanimate objects like the Sun and Moon and anything else that had mass exerted invisible forces on each other that could span vast distances of empty space and act instantaneously. His invoking of gravity as an innate force that was just there and did not have a mechanical explanation was reflective of an earlier way of thinking that had been rejected by his contemporaries and was seen by them as a reversion to older, borderline heretical ideas that attributed occult properties to matter (Butterfield 1957, 157). That would have been a difficult idea to stomach and indeed Newton was accused by Gottfried Wilhelm Leibniz of abandoning the goal of specifying a completely mechanical universe and reintroducing occult principles (Shapin 1996, 63). By saying that he had no choice in the matter and that he was forced to his conclusions by the data, Newton could deflect the charge that he was indulging in mere hypothetical speculations or the occult and claim that his knowledge was certain, thus implying infallibility and inevitability.

Even if Newton's assertions of the inevitability and thus uniqueness of his theories were merely a rhetorical strategy on his part to overcome opposition to his theory, he would not be the last to adopt it. Andre Ampere (1775–1836) said in his treatise on electricity that he had followed Newton's example and that his own laws were also deduced only from observed facts. Here are Ampere's own words as quoted by Duhem:

To observe the facts first, to vary their circumstances as far as possible, to make precise measurements along with this first task in order to deduce from them general laws based only on experience, and to deduce from these laws, independently of any hypothesis about the nature of the forces producing the phenomena, the mathematical value of these forces, i.e., the formula representing them—that is the course Newton followed. It has been generally adopted in France by the scientists to whom physics owes the enormous progress it has made in recent times, and it has served me as a guide in all my research on electrodynamic phenomena. I have consulted only experience in order to establish the laws of these phenomena, and I have deduced from them the formula which can only represent the forces to which they are due; I have made no investigation about the cause itself assignable to these forces, well convinced that any investigation of this kind should be preceded simply by experimental knowledge of the laws and of the determination, deduced solely from these laws, of the value of the elementary force. (Duhem 1906, 196)

After these stirring words to the power of experience to uniquely determine his mathematical laws of electrodynamics, Ampere frankly admits at the end of this *same treatise* that he had not conducted some of the experiments that he just said he had used to derive his laws and indeed that he had not as yet even built the equipment that would have enabled him to do so! Duhem quotes Ampere, saying "I think I ought to remark in finishing this memoir that I have not yet had the time to construct the instruments represented in Diagram 4 of the first plate and in Diagram 20 of the second plate. The experiments for which they were intended have not yet been done" (Duhem 1906, 198–99). As Duhem comments, "Very far from its being the case that Ampere's electrodynamic theory was *entirely deduced from experiment*, experiment played a very feeble role in its formation: it was merely the occasion which awakened the intuition of this physicist of genius, and his intuition did the rest" (199, emphasis in original). Similarly, it is not clear that Galileo did the many experiments of balls rolling down inclined planes that he said he did that showed strong support for his new dynamical theories (Shapin 1996, 84). And as we will see later, Niels Bohr, Werner Heisenberg, and others used similar rhetoric of experimental inevitability to overcome resistance to their favored Copenhagen interpretation of quantum mechanics. Even late into the twentieth century, this idea of inevitability continues to be promoted, as expressed by cosmologist D. W. Sciama when he wrote in the preface of his book that "Newton's laws of motion are logically incomplete by themselves, and the problems they raise lead one step by step to the full complexity of General Relativity. No arbitrariness appears at any point; each step is forced" (Sciama 1969).

But just as Newton's and Ampere's theories were not forced upon them by the data, neither are any of our present day theories similarly forced on us. Indeed the idea of inevitability has its own Achilles heel. If Newton's theories were inevitable, how could Einstein's theory have replaced it? Yet the idea that theory choice is not forced upon us by data seems to fly in the face of all that we learn about the progress of science. Historical accounts are replete with stories of two rival theories competing for dominance and of crucial experiments determining which one emerged victorious. I have mentioned one such case concerning the particle and wave theories of light. How can that be if we can never eliminate all the potential rivals?

The answer is that while it may *seem* like we have only a limited range of possible theories to work with, that is an illusion. As we shall see later in this book, the seemingly limited range of theory choices available at any given time is the consequence of a very long sequence of judgments that have been made through history that results in giving us what seems like a very limited set of alternative theories to choose from and hence enable data to help us decide between them. There are good practical reasons for wearing these blinkers of historical contingent factors that restrict our vision, and treating the resulting limited choices as the only possible ones. It allows us to portray the evolution of science as one of relentless progress and is, paradoxically, one of the sources of science's strength and success. But if freed from constraints imposed by the past and given a set of data and complete freedom to construct theories to explain them, scientists can come up with an infinite number.

Kuhn explained how this limited number of theories comes about when he argued that what we call modern science emerged along with the idea of a dominant paradigm. This crucial stage resulted in the multiple schools of thought that used to operate largely independently of one another being replaced by a dominant one within which most scientists worked. Once a single paradigm took hold in a field, subsequent development consisted of the dominant paradigm being placed in narrow competition with one rival theory after another in succession. In other words, the search for the "correct" scientific theory is not like (say) a tennis tournament in which a large number of entrants start out on an equal footing and by a process of elimination one emerges as the winner. Instead it is more like chess or boxing world championship matches where there is an existing champion and one rival is selected to challenge the champion. If the challenger loses the title bout, the champion is seen as stronger than before. If the challenger wins, then we have a new champion that others jockey to challenge. But at any given time, there is only the champion and a single challenger for the title.

Just as the public remembers only the epic sporting challenges, so it is with science. Textbooks record only those cases where a new theory successfully challenged the old and the so-called falsifying event is used as the explanation for how the winner was decided, with the scientific community that works in the area of the paradigm in question serving as the referee. In contests such as in chess and boxing, on rare occasions a defeated champion might regain the crown later but in science, as far as I am aware, it has never happened that a once-discredited paradigm regains its dominant position, except perhaps in a highly modified form. The closest is the case of light that was once thought of as a particle, then a wave, and now as having both properties in what we refer to as wave-particle duality within the framework of quantum mechanics.

But the problem of paradigm choice does not go away even in situations where there are only two credible competing theories vying for acceptance because even then the data available do not enable one to make an objective choice. Suppose that in some area of science, one theory explains one set of experimental results while another competing theory explains a second set of results. Suppose further that the two sets of experimental data have a partial, but not complete, overlap in that there is a common pool of data that both theories agree with but the remainder are explained only by one or the other theory but not both. How would you judge which theory was the correct, or at least the better, one? Is it the one that explains more results? Although this criterion superficially seems more objective (since it involves comparing two numbers), a little reflection will show that it is untenable for the same reason that assigning probabilities to theories is untenable. It is relatively easy to do two experiments that differ minutely from each other. In the case of Newton's laws of motion we could, if we wished, repeat experiments an enormous number of times, each with slightly differing masses and slightly differing forces. Should each experiment be counted separately? Or should they be counted as one experiment? How different must two experiments be for them to be considered different? Or take the case of evolution. Suppose that evolutionary theory predicts the existence of a certain kind of species earlier in time. If such a fossil is found, that adds to the credibility of the theory. But is the number of fossils that are found important? Suppose another similar, but slightly different, fossil is found nearby. Does the second discovery make the theory "truer" in some sense? Does it matter if the second fossil discovery is discovered far away from the first? How does one judge whether new facts are adding significantly new support for a theory? There are no obvious answers to these questions and so choosing between competing theories purely on the basis of the number of confirmatory results they produce is not a very satisfactory procedure.

Suppose we reject the idea of using the number of confirmatory results as the critical factor but instead use quality. Suppose we choose the theory that gives *better* explanations to the more *crucial* experiments. This approach seems more reasonable than crudely counting numbers, and indeed is often the one that is used in practice, but its subjective elements are obvious. How do we decide if an experiment is crucial? And how do we decide which results are better? What criteria should we use to make these judgments?

Clearly there is no answer to any of these questions that can be defended on purely objective grounds. They all contain elements of subjectivity. While we could appeal to common sense and good taste to justify our belief that a given theory is true or at least better than its competitors, this hardly constitutes a rigorous proof. To claim that scientific theories reveal the truth about nature and then to concede that we do not really know how to objectively defend the validity of that claim is unpalatable to scientists. But it is the use of *scientific logic*, to be discussed in Part Four, that enables us to have such a great degree of confidence in our theories even in the absence of objective certainty.

(e) Scientific theories must be naturalistic and testable

The immense success of science in creating useful and reliable knowledge has led to a situation in which attaching the label of science to a body of knowledge gives it a cachet, bestowing on it prestige and authority that it might otherwise lack. This has led in recent times to the creation of phony "scientific" journals and conferences that lack any of the normal quality controls but only require payment of a fee for papers to be accepted by them. These "papers" are then cited as scientific evidence for all manner of dubious claims (Burdick 2017; Alecci 2018).

But what makes something genuinely science? As I noted earlier, the search for criteria that can distinguish between science and nonscience has come to be known as the *demarcation problem* and the Holy Grail has been to identify a set of *necessary* and *sufficient* conditions that would enable one to say definitively, when confronted with some theory or belief structure, that it is science (if it meets the criteria for sufficiency) or not science (if it fails to meet at least one of the necessary criteria). But despite determined efforts, that search has proved to be futile so far, leading some philosophers of science to argue that the problem of finding clear markers that distinguish science from nonscience may be inherently insoluble (Laudan 1983). This has not prevented others from trying to find ways to distinguish at least the more extreme kinds of pseudoscience from claiming the mantle of science (Pigliucci 2010). But while some epistemologists continue to try to find both necessary and sufficient conditions that would constitute demarcation criteria, in actual practice scientists expect any scientific theory to satisfy at least two necessary conditions, that they be *naturalistic* and *testable*.

Naturalism implies that no appeals to the supernatural are allowed when it comes to seeking explanations for phenomena. This sentiment is captured well by the paleontologist George Gaylord Simpson, who said, "The progress of knowledge rigidly requires that no nonphysical postulate ever be admitted in connection with the study of physical phenomena. We do not know what is and what is not explicable in physical terms, and the researcher who is seeking explanations must seek physical explanations only" (Simpson 1944, 76).

Simpson's formulation may more properly be called *methodological naturalism* and is a *minimal* requirement for scientific theories. He does not rule out a priori the possibility of supernatural phenomena, but says that such things must not be allowed to enter into scientific investigations. Scientists always look for natural explanations to the phenomena they encounter because such an approach allows you to systematically investigate open questions and not shut off avenues of research. Any scientist who said that an experimental result was due to gods or spirits intervening in the laboratory would be looked at askance, because that scientist would be violating one of the fundamental rules of operation. There is no question in science that is closed to further investigation of deeper natural causes. The cartoon by Sidney Harris that can be found pinned to the walls of the offices of many scientists makes the same point humorously (Figure 10.1).

Pretty much everyone, scientists and nonscientists alike, practices this form of methodological naturalism because life would be a nightmare without doing so. For example, if you hear a strange noise in the next room, you might wonder if it is a radiator or the wind or a mouse or an intruder and you investigate each possible natural cause, looking for evidence. You would be unlikely to say, "The noise in the next room is caused by ghosts knocking over stuff." In general, people don't routinely invoke the supernatural to explain the everyday phenomena of their

Figure 10.1 Credit: SCIENCECARTOONSPLUS.COM

lives, even if they are quite religious and superstitious. Methodological naturalism is just that same idea.

Nonscientists sometimes do not understand how hard and frustrating much of scientific research is. In trying to solve tough problems, scientists work for long periods of time banging their heads against what must seem like unyielding walls. What keeps them going? What makes them persevere? It is methodological naturalism, the belief that a discoverable explanation *must* exist. This is what drives medical researchers to work for years, and often decades, to find causes (and thus possibly cures) for diseases. Part of the reason for this doggedness is the desire to be helpful, part of it is due to personal ambition and career advancement, but an important part is also *the belief that a solution exists that lies within their grasp* and that given sufficient ingenuity and skill and time and resources and hard work, they will find the solution. It is because of this willingness to persevere in the face of enormous difficulty that science has been able to make all its tremendous breakthroughs.

Unsolved problems are seen as challenges to the abilities of individual scientists and the scientific community, not as manifestations of supernatural phenomena outside the bounds of science. If, as a result of difficulty in solving a problem, scientists threw up their hands and said, "Well, it looks like supernatural forces are causing this one. Let's give up and move on to something else," then the great discoveries of science that we associate with Newton, Darwin, Einstein, Planck, Heisenberg, and so on might never have occurred. As biologist Richard C. Lewontin said, "We cannot live simultaneously in a world of natural causation and of miracles, for if one miracle can occur, there is no limit" (Lewontin 1983, xxvi). This is why there was such strong opposition from the overwhelming majority of the scientific community, including even those who are religious, at attempts to introduce into school science curricula supernatural ideas such as intelligent design as alternatives to evolution. Such a move was seen as violating a cardinal principle of scientific practice.

There is a stronger formulation of naturalism that is referred to as *philosophical naturalism*. This is the belief that the material world governed by natural laws is all there is and no supernatural phenomena exist at all. The absence of supernatural forces cannot, of course, be *proven* for all the problems that have been discussed earlier with proving a negative. But as we have seen, science is not in the business of proving things to be true or disproving them either. Science is in the business of figuring out what works best in any given situation, using the logical and evidentiary methods that it has found useful. And in that endeavor, appeals to gods or the supernatural are not helpful and hence play no role.

Given this approach in their scientific practice, one should not be surprised to find greater levels of skepticism of the supernatural among scientists than in the general public. It is not the case that all scientists are atheists. Scientists span the entire religious spectrum, from fundamentalist believers through moderate religionists to agnostics and atheists, and as one goes back in time, one finds an increasing number of eminent scientists who were quite religious. Within the world of physics alone, Isaac Newton, James Clerk Maxwell, and Michael Faraday, to name a few, were eminent scientists who were religious in varying degrees. But they usually eschewed any attempt to insert supernatural ideas into explanations of phenomena and were scorned by their fellow scientists if they attempted to do so. We saw an example of this in subsection 10(b) in the derisive responses of his contemporaries to Newton's suggestion of a cosmic reset button periodically pressed by the deity to explain the stability of the solar system. Some of the criticisms of Newton came from scientists who were religious like him, because Newton's suggestion seemed to imply that God was an incompetent engineer who could not foresee that the universe he was creating and setting in motion was going to malfunction in the future (Butterfield 1957, 125).

The *practice* of modern science can undoubtedly be labeled as atheistic in that it does not allow for any supernatural intervention, and there are those scientists who, impressed by the success of eliminating the supernatural as an explanatory concept within their own sphere of science, see no reason why that should not be extended to *all* areas of life, and thus they adopt philosophical naturalism. As population biologist J. B. S. Haldane explained, "My practice as a scientist is atheistic. That is to say, when I set up an experiment I assume that no god, angel or devil is going to interfere with its course; and this assumption has been justified by such success as I have achieved in my professional career. I should therefore be intellectually dishonest if I were not also atheistic in the affairs of the world" (Haldane 1934, vi–vii). The increasing prevalence of this type of reasoning, especially as scientific knowledge grows and successfully fills in the gaps that formerly existed in our understanding of the world, may be why nowadays the proportion of disbelievers is so much higher within the scientific community than it is in the general populace.

In fact, the more successful scientists become, as measured by the level of prestige and honors they acquire during their careers, the less likely they are to be religious. Back in 1914, psychologist James H. Leuba did a study in which he found that the level of disbelief or doubt in the existence of God was much higher among scientists (58%) than in the general population, rising even higher to 70% when one looked at those whom he defined as "greater scientists." Those figures rose yet higher to 67% and 85% respectively when Leuba repeated the study two decades later.

In 1996, researchers Edward J. Larson and Larry Witham repeated that study and surveyed religious beliefs among members of the highly prestigious National Academy of Sciences. They now found almost nonexistent rates of belief.

Our survey found *near universal rejection of the transcendent by NAS natural scientists.* Disbelief in God and immortality among NAS biological scientists was 65.2% and 69.0%, respectively, and among NAS physical scientists it was 79.0% and 76.3%. *Most of the rest were agnostics on both issues, with few believers.* We found the highest percentage of belief among NAS mathematicians (14.3% in God, 15.0% in immortality). Biological scientists had the lowest rate of belief (5.5% in God, 7.1% in immortality), with physicists and astronomers slightly higher (7.5% in God, 7.5% in immortality). (Larson and Witham 1998, my emphasis)

In addition to being naturalistic, another necessary feature for any theory to be considered scientific is that it must be *testable*. By that I mean that it must make predictions that can be compared with experiment or observations. As has been discussed in sections 10(a) and 10(b), the fact that a prediction agrees with the data does not necessarily mean that the theory is true neither does failure to agree mean that the theory is falsified. If scientific theories cannot be proven to be true or false, then why test theories? The point is that it is the results of such comparisons of predictions with data that constitute *evidence* in science and, as will be discussed in chapter 18, it is the accumulation of a *body* of such evidence, not any single instance, that is weighed in arriving at *judgments* as to whether theories are true or false or which is the better theory. A theory that makes no predictions that can be tested is of no use to science and will never be considered part of it because it will have no evidentiary basis, however well it purports to explain a given set of phenomena.

There are many ways to generate predictions. The most minimal is if a set of phenomena seems to indicate a pattern of some kind and that pattern is assumed to apply to all similar classes of phenomena. So for example, after Kepler showed that all the known planets in the solar system moved in elliptical orbits, a reasonable prediction would have been that any new planet that was discovered should also move in an ellipse, as would the planets in any new planetary system that might be discovered elsewhere in the universe. This kind of prediction is based on a form of inductive reasoning.

Another way that predictions can be made is using correlations. If two phenomena are seen to behave in a coordinated way, that correlation can be used to generate predictions. For example, take the suggested connection between lead and violent crime. It has been argued that lead in the environment of young children as they grow up can lead to an increase in their propensity for violence when they become adults, because early lead exposure increases the likelihood for impulsivity, aggressive behavior, and low IQ. The basis for this argument is correlational, in that various parts of the world took steps to eliminate lead in paints and gasoline at different times, and reductions in violence were seen about two decades after those measures were taken (Nevin 2007; Reyes 2007). Such correlations can be used to make predictions in other situations where the pattern should be expected to hold. Since we cannot do experiments that control for lead in humans, these correlations are all that we have so that the argument for a causal relationship has to be circumstantial. But inferring causality from correlations can be justified under certain conditions (Hill 1965).

The most desirable kinds of predictions are those that are based on a proposed theory or law or mechanism for some phenomenon. While the predictions based on Kepler's laws involved using only patterns, and those for lead involved correlations, Newton's laws of motion and gravity not only explained Kepler's laws but went further and explained *why* the orbits had to be ellipses and allowed one to make far more detailed predictions of planetary motions.

This condition of testability based on predictions is the reason why, for example, the idea of intelligent design was never taken seriously by the scientific community as a viable alternative to the theory of evolution by natural selection. Intelligent design advocates limited themselves to merely identifying instances that they felt evolutionary theory could not adequately explain (Johnson 1991; Behe 1996; Wells 2000). They did not take the next step and suggest a pattern or correlation or mechanism that could be tested by telling researchers where to look in order to generate the required evidence in support of their theory. After pointing out what they felt were the failures of evolutionary theory, they pretty much rested their case. Their own explanation for those supposedly inexplicable phenomena were similar to those of the scientist in the Harris cartoon who inserts "Then a miracle occurs" into his chain of reasoning. Such claims will never meet the criteria for being considered part of science.

The advocates of intelligent design seemed to be basing their case on the erroneous belief that there were only two alternatives, evolution or intelligent design, and that exposing the weaknesses of the former would mean that the latter wins by default. As we have seen, it has long been established that that is not how science works. Those who, for whatever reason, dislike the theory of evolution (or naturalistic theories of the origins of the universe) actively seek out unsolved problems in those areas and claim that this proves those theories to be false and thus their preferred myth must be true since those are the only two alternatives. Such people display a profound misunderstanding of the nature of science because, as Duhem said, "the truth of a physical theory is not decided by heads or tails." There are *never* only two theories at play and to command acceptance, a theory must provide a preponderance of *positive* evidence in its favor, not just negative evidence against a competitor. What is more, even their claims about the failures of evolutionary theory have been contradicted by biologists and philosophers (some of whom were religious) as not being valid (Miller 1999; Morris 1998; Shanks 2004).

So to sum up, although all scientific theories and laws are provisional in nature in that we cannot prove them to be true nor can we prove them to be false, and although no amount of data can uniquely determine a theory or law, being naturalistic and having the ability to make predictions that can be tested is essential for any scientific theory because of the crucial role that evidence plays in arriving at *judgments* as to what theories we consider to be provisionally true. Requiring testability ensures the generation of data that can serve as evidence and is what makes scientific theories empirical.

Subsequent chapters will discuss how we use that evidence within the framework of scientific logic. But in order to understand how scientific logic works, we need to first look at the deep interconnectedness of scientific theories in general, something we already encountered in answering the specific question of the age of the Earth.

11

The deep interconnectedness
of scientific theories

[This chapter expands upon the point touched on in the previous chapter of how scientific theories are so deeply interconnected that they cannot be investigated in isolation, and how this prevents individual theories from unequivocally being proven true or false and creates difficulties when choosing between two competing theories.]

Many people believe that the scientific process involves two distinct and independent entities, scientific theories on the one hand and experimental or observational results on the other, and that these can be compared with each other. Theories make predictions for what we should see and experiments check if that is indeed what we see. If they agree, we say that the theories are likely to be true. If they disagree, we say that the theories are false. Hence if a theory sticks around for some time, people may think that this is because no observations contradict it in any serious way and thus it is likely to be true. This highly simplified view leads them to not appreciate the serious problems that confront attempts at finding true theories, some of which were discussed in the previous chapter.

One fundamental problem that arises in determining the truth and falsity of scientific theories is that there is no *single* theory that encompasses all of science. Instead, what we have is a patchwork of theories for different areas of science and *these theories are inextricably interconnected with one another*. A scientist investigating a question of interest does not have the luxury of isolating one theory for study but has to work within a framework of many theories that can range far and wide across many disciplines. If this entire theoretical system produces a result that is consistent with the data, then we can say that there is no problem. But if there is disagreement, then clearly there is a problem *somewhere* even though the source of it may not be immediately obvious and requires careful teasing out.

Each scientist is usually a specialist in one field and will first assume that the theory in his own domain is the possible culprit and will try to probe that theory since it falls within his own area of expertise and is where he has the most competence. All the other theories that have implications for his work but are outside his field are assumed to be well tested and reliable. If reasonable modifications to the suspected theory get rid of the discrepancy and results in bringing the entire system back into harmony, then the scientist can feel confident that he has acted correctly and that the theory was indeed at fault. But one can never be certain that the theory is now true. A new discrepancy may arise elsewhere that is more stubborn and remain unsolved despite all conceivable modifications to the theory under question. In such a situation, should one discard the theory being probed or should one assume that the problem lies further afield, somewhere among all the other theories that impinge on this area? There is no easy answer to this question.

This interconnectedness of scientific theories is similar to the way that all the organs and systems of the human body are interconnected so that the working of each part is influenced by every other part, however unrelated it may superficially appear. The philosopher of science Pierre Duhem analogized this concept over a century ago, saying that when investigating the truth of a particular scientific theory, the scientist acts less like a watchmaker and more like a physician. If we take a malfunctioning watch to a repair shop to figure out what is wrong, the watchmaker can detach each component of the watch from the rest of the machinery and independently test it to see if it is functioning properly. If it is, then he checks another component. By this method of *independent* tests conducted *sequentially*, the watchmaker can systematically proceed until he identifies the exact source of the problem and fixes it. But a physician faced with an ill patient does not have the luxury of removing and isolating each component of the body for independent testing but has to examine each one in situ.

Here are Duhem's own words.

People generally think that each one of the hypotheses employed in physics can be taken in isolation, checked by experiment, and then, when many varied tests have established its validity, given a definitive place in the system of physics. In reality, this is not the case. Physics is not a machine which lets itself be taken apart; we cannot try each piece in isolation and, in order to adjust it, wait until its solidity has been carefully checked. Physical science is a system that must be taken as a whole; it is an organism in which one part cannot be made to function except when the parts that are most remote from it are called into play, some more so than others, but all to some degree. If something goes wrong, if some discomfort is felt in the functioning of the organism, the physicist will have to ferret out through its effect on the entire system which organ needs to be remedied or modified without the possibility of isolating this organ and examining it apart. The watchmaker to whom you give a watch that has stopped separates all the wheelworks and examines them one by one until he finds the part that is defective or broken. The doctor to whom a patient appears cannot dissect him in order to establish his diagnosis; he has to guess the seat and cause of the ailment solely by inspecting disorders affecting the whole body. Now, the physicist concerned with remedying a limping theory resembles the doctor and not the watchmaker. (Duhem 1906, 187–88)

He then states what this implies for the testing of scientific theories.

> To seek to separate each of the hypotheses of theoretical physics from the other assumptions on which this science rests in order to subject it in isolation to observational test is to pursue a chimera; for the realization and interpretation of no matter what experiment in physics imply adherence to a whole set of theoretical propositions.
>
> The only experimental check on a physical theory which is not illogical consists in comparing the *entire system of the physical theory with the whole group of experimental laws*, and in judging whether the latter is represented by the former in a satisfactory manner. (Duhem 1906, 199–200, emphasis in original)

Duhem is arguing that one can never deduce with certainty whether any individual scientific theory is false, *even in principle*. What is now known as the Duhem-Quine thesis states that it is impossible to test a scientific theory in isolation because of the inevitable presence of various background assumptions and auxiliary hypotheses. But anyone with even a cursory knowledge of scientific history knows that individual scientific theories have routinely been pronounced wrong and been replaced by new ones. How could this happen if we cannot isolate a single theory for comparison with data? How can scientists decide which of two competing theories is better at explaining data if a whole slew of other theories are also involved in the process? Is Duhem saying that we can *never* arrive at any conclusion about the truth or falsity of any scientific theory?

Not quite. We saw earlier that Duhem appealed to "good sense" based on experience that enables a scientist to figure out which theory in the entire ensemble is most likely to be wrong and how to fix it. He said that like a physician, a scientist has to exercise a certain amount of discerning judgment in identifying the possible source of the problem, and after conducting tests on the whole system and weighing the evidence, can zero in on one part as the *most likely* source of the problem, all the while being aware that *one does not know for certain*.

Duhem's appeal to the good sense of individual scientists and of the collective scientific community might seem unsatisfactory to many. It seems to be a thin reed on which to base our beliefs in the truth of theories given our heavy dependence on them. Given the tremendous success of science, surely there must be a more definitive method that scientists use to pinpoint the exact source of the problem? Historians and philosophers of science have searched diligently to try to identify features of scientific practice that would be objective and provide the kinds of guarantees that we seek that scientific theories that work well are true. They have failed. And yet over time the scientific community has developed a set of implicit and explicit methods and procedures that enable them to arrive at a consensus on what is true and what is false that flesh out what Duhem vaguely called good sense.

These ways of using logic and evidence and reasoning have proven to be so powerful in arriving at scientific theories that work well that they enable us to have great confidence in those theories, so that even if we cannot *prove* them to be true, to deny their validity would be to act with a *lack* of good sense. These methods have been so successful that, though we may not be consciously aware that we are doing so, they have permeated everywhere and become ubiquitous and we now use them routinely in our everyday lives as well, such as when we accept that smoking increases the risks of getting cancer or reject the existence of werewolves. So when some assert that the scientific consensus in some area should be ignored, *they are rejecting the very methods of reasoning based on scientific logic that they use in their everyday lives* and thus are effectively throwing the proverbial baby out with the bathwater.

To understand the nature of scientific logic, we first begin with how theories get invented and later accepted.

12

How scientific theories get invented and that history gets distorted

[This chapter looks at how we have a natural propensity to invent theories all the time based on our experiences, and it is the testing of these theories and their refutation and replacement with new theories that more accurately represents scientific practice.]

It is important to realize that there is no such thing as *the* "scientific method" in the sense of a single technique or way of operation by scientists. In fact, any broad generalization about how scientists operate can be challenged and when I make seemingly sweeping statements of that nature, it should be understood that I am only implying a widespread prevalence and not unanimity. Students may learn early in their science courses and when doing science projects that they should start with a hypothesis, then collect data and other forms of evidence, and then judge whether they confirm or disprove the hypothesis. In real life, science is a lot messier and involves many possible paths. A new investigation is often triggered when a scientist notices something that gets her attention and intrigues her, maybe when investigating something else. Sometimes it is because what she sees is not what she expected. Or an idea strikes her while idly thinking. Usually the trigger is something odd that does not quite fit into the pattern of existing knowledge or the idea takes the form of perceiving an underlying connection or pattern in what on the surface may seem like unconnected facts. If intrigued enough and she has the time and resources, she may start looking into it more closely to see if there is anything substantive there. This stage of the process can best be described as "messing around," testing out one ad hoc idea after another, trying this and discarding that. If such tentative efforts provide some promising avenues of study, she may embark on a more deliberate research program. In the course of this investigative process the original idea will undergo many revisions and may often be unrecognizable when presented in the final product of a research paper.

Karl Popper pointed out that the relationship between theories and data is dialectical. Scientists often start by postulating a tentative theory to explain a pattern in some set of data. In the process of investigating this, they will often generate new data themselves or learn about the existence of data generated by others that requires modifying their initial theories in order to incorporate this new information. These new theories will then make new predictions that require further testing, which generates yet more data. If the tests start to work out, that is a sign that the scientist may be on the right track. But if the tests yield discrepant results, the scientist will try to modify one of the theories to bring experiment and theory back into agreement. If that can be done without being purely ad hoc, then the revised theory becomes the basis for further testing. This dialectical process between theory and experiment continues until the scientist is satisfied that sufficient agreement between the two has been reached. Then the work is written up and submitted for peer review and publication, and this process may require yet more refinements of the theory.

This process of trial and error forms a large part of everyday scientific work. Popper asserted that all human beings, not just scientists, are *pattern seeking individuals* who are born with an *innate* tendency to make conjectures, to construct a universal theory based on whatever data is at hand, however meager, and to hold on to that theory until it is refuted by new data, whereupon it is immediately replaced with a new universal theory. One does not need multiple repetitions of events in order to generate a belief. Even a single event, if it carries sufficient impact, can be enough. A child who is burned even once by touching a hot stove needs no further tests to be wary of doing so again. This process of conjectures and refutations goes on all the time and explains how science functions. He claimed that this model also solved the *problem of induction* because it explained why we expect that things that have always happened in the past will continue to happen in the future, when logically there is no reason to think so. Popper said that we cannot help it. We have simply evolved to be that way.

> Instead of explaining our propensity to expect regularities as the result of repetition, I proposed to explain repetition-for-us as the result of our propensity to expect regularities and to search for them.
>
> Thus I was led by purely logical considerations to replace the psychological theory of induction by the following view. Without waiting, passively, for repetitions to impress or impose regularities upon us, we actively try to impose regularities upon the world. We try to discover similarities in it, and to interpret it in terms of laws invented by us. Without waiting for premises we jump to conclusions. These may have to be discarded later, should observation show that they are wrong.
>
> This was a theory of trial and error—of *conjectures and refutations*. It made it possible to understand why our attempts to force interpretations upon the world were logically prior to the observations of similarities. Since there were logical reasons behind this procedure, I thought that it would apply in the field of science also; that scientific theories were not the digest of observations, but that they were inventions—conjectures boldly put forward for trial, to be

eliminated if they clashed with observations; with observations which were rarely accidental but as a rule undertaken with the definite intention of testing a theory by obtaining, if possible, a decisive refutation.

The belief that science proceeds from observation to theory is still so widely and so firmly held that my denial of it is often met with incredulity. I have even been suspected of being insincere—of denying what nobody in his senses can doubt.

But in fact the belief that we can start with pure observations alone, without anything in the nature of a theory, is absurd; as may be illustrated by the story of a man who dedicated his life to natural science, wrote down everything he could observe, and bequeathed his priceless collection of observations to the Royal Society to be used as inductive evidence. This story should show us that though beetles may profitably be collected, observations may not.

Twenty five years ago I tried to bring home the same point to a group of physics students in Vienna by beginning a lecture with the following instructions: "Take a pencil and paper; carefully observe, and write down what you have observed!" They asked, of course, *what* I wanted them to observe. Clearly the instruction, "Observe!" is absurd . . . Observation is always selective. It needs a chosen object, a definite task, an interest, a point of view, a problem. (Popper 1965, 46, emphasis in original)

When theory and experiment disagree, it is usually theory that is seen as the problem because, after all, theories are seen as human inventions while experimental results are seen as what an implacable nature is telling us. Good scientific theories are often robust and flexible enough to accommodate initially discrepant data. But sometimes faith in the rightness of the theory is so strong or the changes required to get agreement would so deeply undermine the essence of the theory that the experimental results become suspect and force a reexamination of the data. In 1919, after Arthur Eddington had confirmed the prediction of general relativity that the path of light waves could be bent by the gravitational force of a large object like the Sun, Einstein was asked by a doctoral student what he would have done if the experiment had disagreed with his prediction. Einstein is said to have replied "Then I would feel sorry for the good Lord. The theory is correct anyway" (Calaprice 2005, 226). But not all proposers of a new theory have the confidence of an Einstein.

The conventions of science literature require that when the end result of a research process is finally submitted for publication, the scientist will *reconstruct the narrative* to eliminate all the muddling and false trails that result from the trial and error process, and suggest that it all started with an initial hypothesis that framed the research and final conclusions. This is a false picture in a strictly historical sense but is not intended to deceive because all scientists know that this imposed narrative structure is simply following the conventions of scientific reporting (Medawar 1964). Part of the reason is economy. The scientific literature is dense and voluminous enough as it is. To add all the false trails in published papers would be to make the problem even worse. Furthermore, there is a strong belief among scientists that we are proceeding toward truth. If what science reveals are objective truths about nature, then we believe that we were bound to get there eventually and *how* we got there is relatively unimportant. When it comes to theories that we believe are right, it is the destination and not the journey that matters for the scientist, with the journey seen as important only for the historians of science who reconstruct it from the papers, talks, notes, correspondences, and memoirs of the participants.

But another reason for the truncated narrative in scientific papers is that it is a quirk of the brain that once we have struggled to understand something clearly and succeeded, it is often hard for us to remember our confused initial state. The researchers themselves may forget many of the wrong assumptions they started out with and the wrong leads they followed and remember only those that pointed them toward the final conclusion. Incidentally this is also why it is hard for experts in a field to teach novices, because they find it hard to place themselves in the mindset of the learners they themselves once were and are unable to recall what they found difficult. Everything seems so straightforward to them now that they cannot understand why others are so confused and cannot see what they so clearly see. The truly great teachers are those who realize this feature of the brain and work hard to acquire the ability to place themselves once again in the shoes of the novice.

But the phenomenon of forgetting the initial state extends beyond the individual researcher or the individual teacher and learner and extends to the collective remembrance of the scientific community as a whole. When historians of science look closely at specific episodes of scientific discovery using contemporaneous records, what they find are parallels to what an individual scientist goes through in carrying out research. The process is often confused and chaotic with competing schools of thought and various blind alleys being pursued and successes achieved in fits and starts until finally a consensus emerges as to what are seen as the correct conclusions and the right way of doing things. Often it is hard for the historian to identify a specific point in time when a discovery was made and who was the first to make the discovery, even though priority and credit for discoveries are features that are highly prized in the scientific community (Butterfield 1957).

But when that same history is written down in science textbooks, a completely different picture is drawn. The narrative is reconstructed to provide a clean line of progression, to suggest that theories are created at one time by one person (or a few people), enjoy a period of success, and then are discarded in favor of a new theory when they are found to be falsified due to new evidence being unearthed. This narrative feeds into the perception that scientific theories are getting better and better all the time and heading steadily toward the ultimate goal of arriving at true theories.

As Thomas Kuhn describes, the straightforward question "Who discovered the existence of oxygen and when was it discovered?" has no simple answer. In examining the historical record, he says that "we can safely say that oxygen had not been discovered before 1774, and we would probably also say that it had been discovered by 1777, or shortly thereafter." But in between those dates, the status of oxygen was ambiguous and both Joseph Priestley in England and Antoine Lavoisier in France have plausible claims to being the discoverer. And yet many popular accounts will pick on one particular event and use it to flatly state that either Priestley discovered oxygen in 1774 or that Lavoisier did in 1775, with national pride sometimes factoring into their judgment (Kuhn 1970, 55).

A more recent example of this is the credit for the discovery of the Higgs particle. The idea for the existence of this particle was being bandied around during the 1950s and 1960s by many scientists, at least six of whom had credible claims to have contributed significantly. After the particle was directly detected in 2012, it was obvious that the Nobel prize physics committee would give its prestigious prize to the people who proposed the original idea and immediately a controversy arose, since the rules state that the award must be given only to living scientists and can be shared by at most three recipients. Since five of the six were still living, this created a problem. The prize was ultimately awarded to just two of the five, naturally causing some unhappiness in the community. Since one of the two is Peter Higgs, the scientist whose name has become attached to the particle, it is quite likely that when textbooks and articles are written in the future, when people have forgotten all the complicated history behind the origins of the theory, credit will be assigned to just this scientist and the date will be that of his publication in 1964. This is how scientific history becomes truncated and distorted over time.

Another example is the photoelectric effect. Textbooks state that this effect was inexplicable using classical electromagnetic radiation theory that presumed that light consisted of waves, as was established earlier following the work of Thomas Young, Augustine Fresnel, Dominique Arago, and others. Einstein's paper in 1905 that suggested that light consisted of particle-like entities called photons, a partial reversal to Newtonian ideas, resolved the photoelectric problem and it is for this and not for his better known theories of relativity that he was awarded the Nobel prize in 1921. But in reality, while his paper is now described in textbooks as being decisive in persuading scientists that the photon model is the *only* possible explanation for the experimental results, other eminent scientists had proposed alternative explanations for the photoelectric effect that did not involve photons. The controversy persisted until other experiments were done and it was the *combination of all of them* that persuaded the community of physicists to adopt the photon explanation as the correct one. As Roger Stuewer says, "Certainly nothing is more artificial than concentrating on one experiment to the exclusion of all other phenomena: the photon concept, like all fruitful concepts in physics, derives its validity from an interlocking theoretical and experimental matrix." Stuewer adds that alternative theories exist even today and that statements found in textbooks about the inevitability of the photon model being correct, such as that there "is absolutely no possible way of accounting for photoelectric effects . . . except by adopting the idea of the photon as a sort of particle carrying its full energy and travelling with the velocity of light" are simply wrong, emphasizing once again that no theory is ever uniquely determined by a single experiment, however clear and decisive it may seem (Stuewer 1970).

So why is this complex history so consistently distorted? Again, this is not due to deliberate dishonesty on the part of the authors of textbooks, books that are sometimes referred to as "the graveyards of science" (Beller 1999, 6), though they sometimes do go too far in their efforts to simplify the historical record and portray science as unrelentingly progressive, only being hindered by human error or obtuseness. Pursuing research at the frontiers of science requires hard work and one of the chief purposes of textbooks is to persuade current students who will form the next generation of scientists as to why they should have confidence in the theories they are working so hard at understanding. In order to get them to the frontiers of science as quickly as possible, the pedagogical goal becomes to convince them that current theories are right and this involves showing why the older theories were wrong, and myths and folklore and distorted history are often the easiest ways of achieving that goal.

There are also eminently practical reasons for teaching science this way. The laws of science, even the most seemingly obvious ones, were only arrived at after extended debates, and the process often lasted decades and even centuries before a consensus was reached as to which theory was the best. In standard science courses, there is simply no time to rehearse all those erudite discussions that involved long chains of deep inferential reasoning. If we tried to do that we would be doing a disservice to those scientists, engineers, doctors, technicians, or other professionals who just want to use the latest and presumably best scientific knowledge, by extending their already lengthy education by many years.

Kuhn argues that despite this simplified and highly narrowed vision and narrative structure of science education, it also provides the most efficient way of achieving the goal of teaching students to become researchers.

Without wishing to defend the excessive lengths to which this type of education has occasionally been carried, one cannot help but notice that in general it has been remarkably effective. Of course, it is a narrow and rigid educa-tion, probably more so than any other except perhaps in orthodox theology. But for normal-scientific work, for puzzle-solving within the tradition that the textbooks define, the scientist is almost perfectly equipped. Furthermore, he is well equipped for another task as well—the generation through normal science of significant crises. When they arise, the scientist is not, of course, equally well prepared. Even though prolonged crises are probably reflected in less rigid ed-ucational practice, scientific training is not well designed to produce the man who will easily discover a fresh approach. But as long as somebody appears with a new candidate for a paradigm—usually a young man or one new to the field—the loss due to rigidity accrues only to the individual. Given a generation in which to effect the change, individual rigidity is compatible with a commu-nity that can switch from paradigm to paradigm when the occasion demands. Particularly, it is compatible when that very rigidity provides the community with a sensitive indicator that something has gone wrong. (Kuhn 1970, 165–66)

The history of the age of the Earth is a good example of what Kuhn describes, where the community of scientists switched allegiance to new paradigms quite rapidly despite the rigidity of individual scientists. But something is also lost in burying the complexities of scientific history in the drive for pedagogical effi-ciency. As Einstein said, "There is always a certain charm in tracing the evolution of theories in the original papers; often such study offers deeper insights into the subject matter than the systematic presentation of the final result, polished by the words of many contemporaries" (Calaprice 2005, 230).

PART FOUR

THE NATURE
OF SCIENTIFIC LOGIC

13

Truth in mathematics and science

[This chapter looks at what we can learn from axiomatic systems and the role of proofs in arriving at truths. It discusses why there are limits to what we can prove to be true even in mathematics because we cannot construct a framework that is both *complete* and *consistent* for any nontrivial system, Science is slightly different in that we deal with *quasi-axiomatic* systems with the additional element of experimental data or observations that we can compare with the predictions of theories, but it faces the same problem.]

As Popper describes it, science progresses by making conjectures, then rejecting them on the basis of evidence, and replacing them with new conjectures. Thus science lurches from one potentially and provisionally true theory to the next. While the idea that theories can be definitively falsified because of contradiction with data has been discredited, Thomas Kuhn says that the kind of rigid education that scientists experience results in them having the ability to identify when an existing paradigm is in need of replacement and thus creates the environment for a revolution that puts in place a new paradigm.

So how does the scientific community decide which theories should be assigned the status of provisional truth or controlling paradigm? In seeking to establish the truth of a scientific proposition, scientists use reasoning and logical arguments that share strong similarities with, *but are not identical to*, mathematical and legal reasoning. The former will be discussed in this chapter and the latter in the next. Being aware of the similarities and distinctions is enlightening and also important to avoid claiming scientific justification for statements that are not valid, as often happens when people try to co-opt science in support of their dubious beliefs or to advance anti-science agendas.

The first issue to consider is the relationship between truth and proof, because in everyday language truth and proof are considered to be almost synonymous. The word "proof" is seen as being definitive, and it is always more authoritative to say that we have *proven* something to be true or false. The gold standard of proof comes from mathematics as do many of our intuitive notions about it so it is worthwhile to see how proof works there, what its limitations are when applied even within mathematics, and what further limitations arise when we attempt to transfer those ideas into science.

Within mathematics, Euclidean geometry is the prototypical system that demonstrates the power of proof and serves as a model for all *axiomatic* systems of logic. In such systems, we start with a set of axioms (i.e., a set of basic assumptions expressed as propositions) and a set of logical rules. By applying the rules of logic to the axioms, we arrive at certain conclusions; that is, we *prove* what are called *theorems*. Using those theorems we can prove yet more theorems, creating an entire hierarchy of them, all ultimately resting on the underlying axioms and the rules of logic. If the axiomatic system we construct is supposed to correspond to some system in the physical world (as Euclidean geometry was once believed to represent the space we occupy), do these theorems correspond to true statements about that world? Yes, *but only if the axioms with which we start out are true and the rules of logic that we use are valid.* Those two necessary conditions have to be established independently.

So how does one do that? While we may all be able to agree on the validity of the rules of logic if they are transparent, simple, and straightforward (though there are subtle pitfalls even there), establishing the truth of the axioms is not easy because things that seem to be perfectly straightforward and unambiguous may turn out to be not so. For example, the fifth and final one of Euclid's postulates, known as the parallel postulate, was long thought to be an obviously true statement about the nature of space but later was found to be a feature of only one class of geometries that are applicable on the small scale of our everyday experience and that other classes, what we now refer to as non-Euclidean geometries, play important roles in the large-scale structure of the universe.

Furthermore, even assuming for the moment that one knows that the axioms are true and the rules of logic are valid, there are still problems that must be addressed. For example, how can we know that all the theorems that we can prove correspond *exactly* to all the true statements that exist? Is it possible that there could be true statements that can never be reached by this process, however much we may grow the tree of theorems? How would we even know if there are true statements out there that we have not proved as yet or indeed *cannot* prove? This is known as the problem of *completeness*.

There is also another problem. Since the process of proving theorems is open-ended in that there is no limit to how many we can potentially prove, how can we be sure that if we keep going and proving more and more theorems that we won't eventually prove a new theorem that directly contradicts one that we proved earlier, thus resulting in the absurdity that a statement and its contradiction have both been proven to be true? This is known as the problem of *consistency*.

To address this second problem, we rely upon a fundamental principle of logic that "truth cannot contradict truth," and thus we believe that it can never happen that two true statements contradict each other. This is why establishing the truth of the axioms and using valid rules of logic is so important because that would *guarantee* that the system is consistent, since any theorem that is based on them must be true and thus two theorems can never contradict each other. Conversely, if we ever find that we can prove as theorems both a statement and its negation, then the entire system is inconsistent and this implies that at least one of the axioms must be false (or at least inconsistent with the others) or that an invalid rule of logic has been used.

If, as stated here, the rules of logic that are applicable in a mathematical system are simple and transparent enough, there is usually little disagreement about their validity, and thus a true set of axioms is taken to imply a consistent system of theorems and vice versa. Hence we can at least solve the problem of consistency if we can establish the truth of the axioms, though the completeness problem still remains open.

Those who are familiar with these issues will recognize that we are approaching the terrain known as Gödel's theorems. The key result that Gödel showed is that for any nontrivial system, *it is impossible to prove either completeness or consistency*. In other words, there will be true statements that cannot be proved to be so and that it is possible for a theorem to emerge that contradicts another theorem.[1]

Some have seized upon Gödel's conclusion that true statements exist that cannot be proved to try to make claims of truth for their unprovable beliefs. There have been suggestions that "God exists" is one of those statements. This is a misunderstanding of what Gödel proved but is typical of attempts by people who seize upon and use important results in science and mathematics (especially those that impose some limits on knowledge, such as the uncertainty principle or Gödel's results) to squeeze in things for which there is little or no evidentiary support. Incidentally, Gödel himself proposed an ontological proof of God, but that effort is distinct from his incompleteness theorem (Oppy 2017).

The fact is that we cannot simply insert any proposition we choose into the vacant niches that Gödel discovered. The true yet unprovable statements have to be constructed within that particular axiomatic system and are thus dependent on the axioms used. A statement that is true but unprovable in one system need not be so in another one. Simply by adding a single new axiom to a system, statements that were formerly unprovable cease to be so, while new true unprovable statements emerge. For example, initially Euclid used just the first four of his five postulates, thinking that the fifth one might be provable from the others. Using those four, he was able to prove up to twenty-eight theorems but those did not include the fifth postulate. He had to add on the fifth postulate as an axiom to prove the twenty-ninth theorem and beyond. Although Euclid's work originally appeared around 300 B.C.E., it was not until the nineteenth century that it was

---

[1] I elaborate on these ideas in the Supplementary Materials C section at the end of the book. For those readers seeking to understand it in more depth I can strongly recommend the short monograph *Gödel's Proof* (Nagel and Newman 2001) and the entertaining (but much longer) *Gödel, Escher, Bach: an Eternal Golden Braid* (Hofstadter 1999). For a more rigorous treatment, the reader is referred to *An Introduction to Gödel's Theorems* (Smith 2013).

demonstrated that it was indeed impossible to prove the fifth from the other four, and thus that it was necessary to treat it as an additional assumption.

Whenever people invoke Gödel's theorem (or the uncertainty principle or information theory) in support of their beliefs, we should be on our guard and investigate if what they say is actually what the science or mathematics says. We should be wary of the easy option of ascribing deeper meaning to what may simply be due to our lack of imagination or ingenuity or perseverance. Indeed, it is the steadfast refusal of mathematicians and scientists to concede that there is *any* theorem that lies beyond their ability to prove, or any problem that is beyond their capacity to solve, that has resulted in some of their greatest successes, such as finding a proof of Fermat's Last Theorem.

So what can we do in the face of Gödel's implacable conclusion that we cannot construct an axiomatic system in which the theorems are both complete and consistent? At this point, pure mathematicians and scientists part company. For the former, the truth or falsity of their theorems (and hence of the axioms) is not the only thing of interest. What is also important is whether the conclusions they arrive at (the theorems) are the *necessary conclusions* of their chosen set of axioms and rules of logic. Even a statement such as "1+1=2," which most people might regard as a universal truth that cannot be denied, is seen by them as merely the consequence of certain starting assumptions, and one cannot assign any absolute truth value to it. So pure mathematicians concern themselves more with the *rigor* of proofs, and less with whether the theorems resulting from them have any meaning that could be related to truth in the empirical world. What is important is that the axioms be consistent, or at least appear to be so since we can never prove them to be so. Whether they say anything about the physical world that can be described as true has ceased to be determinative.

For scientists dealing with the empirical world, however, questions of truth remain paramount. It matters greatly to them whether some result or conclusion is true or not. But if truth is so elusive even in the rigorous field of mathematics, what chance do we have of achieving it in the much messier world that constitutes science? While the methods of proof that have been developed in mathematics are used extensively in science, scientists have had to look elsewhere other than proofs to try and establish the truth or falsity of scientific propositions and theories. And that "elsewhere" lies with empirical data or the "real world" as some like to call it, which is why the notion of evidence plays an essential role in science. So while in mathematics the statement "1+1=2" is simply a string of symbols representing a theorem based on a particular set of axioms and rules of logic, in science, its empirical truth or falsity is extremely important and is judged by how well real objects (apples, chairs, etc.) conform to it.

This dependence on data raises a problem similar to that of the consistency and completeness problems in mathematics that Gödel highlighted. We can see if "1+1=2" is true for many sets of objects by bringing in actual objects and counting them. But we obviously cannot do so for *everything* in the universe. So how can we know that this result holds for all objects all the time, that is, that it is a universal truth? Such a concern may well seem manifestly overblown for a simple and transparent assertion like "1+1=2" but many (if not most) results in science are not obviously universally true and so they can be challenged on the grounds that they have not been proven to be true in all cases.

For a long time, it was religion that claimed to reveal eternal and universal truths. No one except true believers seriously says that anymore because science has become the source of reliable knowledge while religion is increasingly seen as being based on evidence-free assertions. So some believers tend to try and devalue the insights science provides by elevating what we can call truth to only those statements that reach the level of mathematical proof, because such a high bar can rarely be attained and thus everything else becomes a matter of opinion. They can then claim that scientific statements and religious statements, and indeed any statements at all, merely reflect the speaker's opinion, nothing more.

It is not just religious people who resort to this type of argument to try to undermine confidence in the conclusions of science. For example, for a long time the tobacco industry challenged the conclusion that smoking causes cancer by pointing out that the link had not been conclusively proven. They argued that there were other possible explanations and pointed out what they claimed was counterevidence, that there exist some smokers who do not get lung cancer and some nonsmokers who do. Nowadays climate change denialists resort to similar arguments to challenge the scientific consensus that human activity is a major factor in causing global warming. So however much the data we obtain supports some proposition, how can we counter those who assert that there might exist some new or existing but yet undiscovered data out there that will refute some scientific claim? Can we ever make definitive statements in science?

Despite the inability to *prove* our conclusions in the mathematical sense of the word, we can be *definitive* in science. The justification of scientific conclusions depends upon a line of reasoning that I have called *scientific logic* that is different from what we think of as involving proofs in mathematics, and understanding it involves a closer examination of the logic system used in science.

In mathematics, the axiomatic method involves starting with a set of axioms (propositions that are self-evidently true or are at least not obviously self-contradictory) and then applying the rules of logic to arrive at theorems. In science the parallel exercise is a *quasi-axiomatic* one. In addition to the mathematical ideas of axioms, logic, and proof, in science we are also dealing with the empirical world and this gives us an additional tool in the form of data for determining the validity of our conclusions. The data usually come either in the form of observations (for those situations that are not replicable, as is often the case in the fields of astronomy, astrophysics, paleontology, evolution, and geology) or in the form of experimental data obtained under repeatable, controlled conditions.

In science we start with a basic theory that consists of a set of entities (particles, organisms, or whatever) and the laws (or principles) and theories that are assumed to apply to them. All of these together serve as the scientific analogues of axioms. It should be emphasized that the basic theories and laws are not conjured up out of nothing. As Popper suggested, they are usually arrived at *inductively*, consisting of generalizations postulated on the basis of observations that seem to show some kind of tentative pattern. That pattern is investigated and successively refined until the basic theoretical framework emerges and we get a well-functioning system. We then apply the rules of scientific logic and the techniques of mathematics to these laws and theories and entities to arrive at conclusions that can be tested.

For example, in quantum mechanics one might start with the Schrödinger equation, the laws of electrodynamics, and a system consisting of a proton and electron each having specific properties (mass, electric charge, and so on). Taking these as the equivalent of "axioms," we then use mathematics to calculate properties of the hydrogen atom, such as its energy levels, emission and absorption spectra, and so on. In biology, one might start with the theory of evolution by natural selection and see how it applies to a given set of entities such as genes, cells, or larger organisms. The results obtained in each case would constitute the analogues of mathematical "theorems," but they are not referred to as such but as *predictions*.

An aside may be necessary about the word "prediction." In everyday language, a prediction usually *precedes* the outcome being predicted. So a successful prediction would be one that had been enunciated *before* an event actually occurred and correctly anticipated the outcome of that event. In science, the word "prediction" is sometimes used *even if the experiment or observation has already been done and the result known*, provided that the knowledge gained by it is not used in creating the theory or in the calculation leading up to the "prediction." Why the word is used by scientists in this somewhat unusual way may be due to the strong belief amongst them that experiment and theory are independent activities and that even if one group knows the result of the other, this prior knowledge of the result will not significantly influence its own activities. For example, theorist Edward Witten asserts that string theory, a relatively recent innovation, "predicts" gravity even though theories of gravity have, of course, been around for centuries. He is quoted as saying "Even though it is, properly speaking, a 'postprediction' in the sense that the experiment was made before the theory, the fact that gravity is a consequence of string theory, to me, is one of the greatest theoretical insights ever" (Horgan 1996; 2014).

We can compare a specific prediction with experimental data and see if the prediction holds up or not. Sometimes this result is very important for its own sake, such as when we are testing to see if a drug or other medical procedure effectively treats a disease. However, what we are often more interested in is the more basic question of whether the underlying theory that was used to arrive at the prediction is true. The real power of science comes from its theories because it is those that determine the framework in which science is practiced and have broad applications. So determining whether a *theory* is true is of prime importance in science, more so than the question of whether any specific prediction is borne out.

It is here that we run into problems with the idea of truth in science. While we may be able to directly measure the properties of the entities that enter into our theory (like the mass and charge of particles), as discussed in chapter 11 we cannot independently test the individual laws and theories that govern the behavior of those particles and show them to be true. Since we cannot treat the basic theories as axioms whose truth can be conclusively established, this means that the predictions we make do not have the status of theorems and so cannot be considered as a priori true. All we have are the consequences of applying the theory to a given set of entities, that is, its predictions, and the comparisons of those predictions with data. *The results of these comparisons are the things that constitute evidence in science.* So in other words, while in mathematics we start by *assuming* a set of axioms to be true and then go *forward* to prove theorems using the rules of logic, in science we work *backward* and try to *infer* the truth of the "axioms" (i.e., the properties of the entities and the laws and theories that govern them) from the degree of agreement that our predictions have with theory.

So what can we infer about the truth or falsity of a theory using such evidence? For example, if we find evidence that agrees with a prediction, does that mean that the theory that predicted it is necessarily true? Conversely, if we find evidence that contradicts a prediction, does that mean that the theory is necessarily false? As I discussed before in chapter 10, it is not that simple. The logic of science does not permit us to make that kind of strong inference. After all, any reasonably sophisticated theory allows for a large (and usually infinite) number of predictions. Only a few of those may be amenable to direct comparison with experiment. The fact that those few agree does not give us the luxury of inferring that all future experiments will also agree, the well-known problem of induction that we have already encountered several times. At best, those successful predictions will serve as evidence in support of our theory and suggest that it is not obviously wrong, but that is about all we can conclude. The greater the preponderance of evidence in support of a theory, the more confident we are about its validity, but we never reach a stage where we can unequivocally assert that a theory has been *proven* to be true. Scientific research is not like a video game where bells ring and flashing screens indicate that one has hit the jackpot and arrived at true theories.

So we arrive at a situation in science that is analogous to that of Gödel's theorem in mathematics, in that the goal of being able to create a system such that we can unequivocally find true theories turns out to be illusory. In the next chapter we will leave behind mathematical systems and discuss how scientific logic compares with that found in legal systems, and see that the use of evidence gives us a method of arriving at firm judgments.

14

The burden of proof in scientific and legal systems

[Scientific logic has strong similarities to the way that the legal system uses logical arguments and evidence to arrive at judgments. The logic used depends on whether a proposition is an existence claim or a universal one, and this determines where the burden of proof lies. That same kind of logic is used also in everyday life, though many people may not consciously realize that they are doing so.]

Since arriving at truth by proof is elusive even in the more rigorous field of mathematics, expecting proof to lead to truth in science is unrealistic. Science needs to use criteria other than proof for making judgments about truth and in doing so, scientists act more like judges in legal cases than mathematicians deriving proofs, and evidence plays a crucial role. For example, in legal proceedings, the usual practice is to follow the principle *ei incumbit probatio qui dicit, non qui negat*, that translates from the Latin as "the burden of proof rests upon him who affirms, not on him who denies," provided the assertion is of a positive nature and not a negative one. This principle is more popularly stated as that a person is *presumed* innocent until shown to be guilty. So if someone is accused of committing a crime, the burden of proof is on the accuser and not the defendant.

This principle is considered such a fundamental aspect of a civilized society that it is enshrined in Article 11 of the Universal Declaration of Human Rights (UDHR), which states: "Everyone charged with a penal offence has the right to be presumed innocent until proved guilty according to law in a public trial at which they have had all the guarantees necessary for their defence." Of course, many countries that have signed on to the UDHR routinely violate this principle when it suits them, while still claiming to uphold the basic principles of human rights.

Although the word "proof" and its derivatives are often used in both legal and scientific contexts, it is not proof in the rigorous mathematical sense but more similar to the words "shown" or "demonstrated." A point to note is that technically the only outcomes in a criminal legal proceeding are "guilty" and "not guilty." *The defendant is never proven to be innocent*, and has no obligation to do so. Indeed the defendant is not even obligated to provide any kind of defense at all and can be exonerated simply because the accuser has failed to make an adequate case. This can of course lead to undesirable situations where the jury can suspect that a defendant is indeed guilty of a crime but feels obliged to bring in a verdict of not guilty if the case has not met the required legal standard, which is why the "proven innocent" phrasing is not appropriate for not guilty verdicts. But this kind of undesirable outcome is the price we pay for trying to have a system that avoids the conviction of innocent people, even if it should lead to outcries of the sort seen following not guilty verdicts in cases where the public strongly believes in the guilt of the accused. This is because the public mistakenly interprets the not guilty verdict as implying that *innocence had been proven* when all it meant was that the *presumption of innocence* had not been sufficiently contradicted.

One could conceive of an alternative system in which a person is presumed guilty until proven innocent, shifting the entire burden of proof onto the defendant. There is nothing *logically* wrong with such a system but in practice it would be unworkable since there are many more people who are innocent of a crime than there are those who are guilty, so one would immediately start with an enormously large pool of presumed guilty people who have to prove their innocence. Furthermore it is often difficult, if not impossible, to establish innocence because it involves proving a negative. For example, if I were alone at home late at night, it would be very difficult for me to prove that I was not robbing a nearby convenient store at that same time, which is why the "presumed innocent until proven guilty" standard is a much better one since it requires at least some evidence that I might be responsible.

Unless one agrees on which of the two frameworks (presumed innocent until proven guilty or presumed guilty until proven innocent) to use in making legal judgments, it may be impossible to agree on a verdict. But whatever system one chooses, the basic reason for doing so is to create a *default position* that is *assumed to be true* unless shown otherwise, so that evidence and arguments are required only to counter the default position.

In science, as in the legal context, there are good reasons for having the burden of proof be on the person who asserts a *positive* claim and not on the person who denies as the method of arriving at judgments. Thus the logic used in arriving at scientific conclusions closely tracks the legal maxim that "the burden of proof rests upon those who affirm," emphasizing once again that the word "proof" used in science does not correspond to the way it is used in mathematics, but more along the lines used in law.

In the legal arena in the United States there are two standards for proof. In criminal cases, there is the higher bar of proving beyond a reasonable doubt, but in civil cases the standard is a lower one based on the *preponderance of evidence*. So if the preponderance of evidence is in favor of one position, it is assumed to be true even if it has not been proven beyond a reasonable doubt. Similarly, scientific logic says that scientific propositions are judged to be true not because they have been proven to be logically and incontrovertibly true (which is impossible to do) or because they have been deemed by knowledgeable experts to be beyond a reasonable doubt (a standard that is not impossible to attain but is often too high a bar to result in productive science except in some rare situations), but because the preponderance of evidence favors them. Evidence plays as crucial a role in science as it does in legal cases when trying to reach such a verdict.

As discussed in the previous chapter, the "axioms" in science consist of the entities that a given system is concerned with and the laws and theories that govern their behavior, all of which are used in making predictions. The evidence that is used in science consists of the comparisons of the predictions of theory with the data obtained by observations or experiments. We then work backward to see if the preponderance of evidence justifies claims that the axioms are true.

Scientific claims can be either *existence* claims (in the case of the constituent entities such as electrons or bacteria) or *universal* claims (in the case of the laws and theories that govern those entities). As mathematician John Allen Paulos points out, these two types of propositions are "proved" in different ways, where the word "proof" is used in the looser legal and scientific sense of being shown or demonstrated by a preponderance of evidence and not in the more rigorous mathematical sense (Paulos 2008, 41–42). In scientific logic, the burden of proof for any existence claim lies, as with legal claims, with those who make the affirmative claim of existence because the default position is that the existence claim is false. If advocates for the existence of any entity cannot meet the accepted standard of proof, the claim is *presumed* to be false even if it cannot be *proven* to be false. Indeed it is often not possible even in principle for an existence proposition to be proven false, but using scientific logic we can still act with confidence that it is false.

For example, the assertion that an entity called an electron exists has to be supported by a preponderance of evidence that an entity with the postulated properties of an electron (such as its mass and charge) has been, or at least can be, detected in experiments. The reason that I say "can be" is that in some cases if there is strong circumstantial evidence in favor of the existence of an entity, a *provisional* verdict in favor of existence may be granted, pending more direct confirmation. A famous case of this is the "luminiferous aether," which, despite the absence of direct evidence for its existence, was postulated to exist on the basis of seemingly strong circumstantial arguments that it *should* exist because there seemed to be no other way of explaining how light could propagate over the vast reaches of space. The aether seemed like a *necessary explanatory concept* and was granted provisional existence status until the theory of relativity undermined the arguments in its favor and its provisional status of existence was then withdrawn. Note that the aether was *not proven to not exist*, because that cannot be done.

On the other hand, the elusive subatomic particle known as the neutrino is an example of something that, like the aether, was also granted provisional existence status and then a quarter of a century later was directly detected. The same is true of the Higgs boson and gravitational waves that were discussed earlier. Because they were also seen as necessary explanatory concepts, physicists believed so strongly in the existence of these highly elusive particles and waves that they were granted provisional existence, and direct confirmation of their existence after long, expensive, and arduous searches was greeted with relief, rather than surprise.

Scientists are willing to grant provisional existence for things even in the absence of any direct evidence in its favor as long as it serves some purpose and can provide tests of existence that they can look for. A good example is the case of magnetic monopoles. For a long time electricity and magnetism were considered to be two distinct aspects of nature though they shared some similarities. They both had two kinds of charge that exerted mathematically similar types of forces. In the case of electricity we called the charges positive and negative while for magnetism we called them north and south poles. But a key difference was that while we could isolate individual electric charges (the proton for a positive charge and the electron for a negative charge), we could not similarly isolate magnetic poles, referred to as monopoles. They seemed to always occur in opposite pairs so that if we cut a magnet in two, we do not get separate north and south poles but instead two smaller magnets each having a pair of poles. However much we keep splitting magnets, even getting down to the basic protons and electrons, we never find an isolated magnetic pole. Since no laws preclude their existence, they could in principle exist.

As long as electricity and magnetism were considered distinct entities, this difference could be shrugged off. But with the discovery that the two phenomena were deeply interconnected in what we now refer to as electromagnetism, the lack of monopoles became more puzzling because the resulting laws of electromagnetism (known as Maxwell's laws) would have an aesthetically pleasing symmetry if magnetic monopoles existed, just like their electric counterparts. Of course, we could dismiss this lack of symmetry as a brute fact of nature. After all, there is no a priori reason why the laws of nature should have any aesthetic character. So the question was whether magnetic monopoles exist but are so rare that we had not seen them as yet or whether they do not exist at all, and if so, whether we could figure out why not.

If magnetic monopoles do exist, it should be possible to look for their experimental signatures, and researchers have long been looking for such indicators. Unfortunately those markers are subtle and can be mimicked by other, more mundane, factors. While there have been reported sightings of monopoles in the past, they have proved to be like sightings of Elvis Presley or the Loch Ness monster, in that they disappear under closer scrutiny. But this has not deterred some physicists from continuing the search because the discovery of monopoles would be major news—not because it would upend the entire world of science (monopoles can be incorporated into Maxwell's laws quite straightforwardly and most areas of science would be unaffected by the discovery)—but because it would restore symmetry to Maxwell's equations and provide some important insights into the fundamental nature of matter by possibly explaining, for example, why electric charges are quantized, that is, occur in little discrete amounts. So while physicists go about their work assuming that monopoles do not exist because there is no evidence for them, they would be delighted if it were shown to be otherwise (Rajantie 2016).

The problem with proving nonexistence of entities is illustrated by the film *Avatar* that postulated the existence of a valuable mineral called Unobtainium on another planet called Pandora somewhere in the universe. Although the film was unabashedly fictional, how could one possibly prove that such a mineral (or even the planet) does not exist somewhere out there? One cannot. Indeed the genre of science fiction routinely assumes the existence of as yet undetected things and universal laws. While such speculations can have great entertainment value and can even spur scientific investigations, the scientific rule is that to establish the existence of some entity, one has to provide a preponderance of *positive* evidence in support of it. In the absence of such evidence, the scientific conclusion is that the existence proposition is false and that the entity does not exist.

This rule is hardly controversial. It is used in everyday life by everyone, even though they may not explicitly recognize that they are doing so. It would be impossible to live otherwise. Small children may fear monsters hiding under their beds but adults know better and do not bother to look and check. The argument that such things *might* exist because we have not proven them to not exist carries no weight at all, and rightly so. To allow for the existence of something in the absence of a preponderance of evidence in support is to open oneself to an infinite number of mythical entities such as zombies, ghosts, unicorns, leprechauns, pixies, dragons, centaurs, mermaids, fairies, demons, vampires, werewolves, and anything else that a fertile imagination can conjure up. Unfortunately, many people do not use this rule *consistently*. They use it *selectively* based on subjective factors, which is why a survey of American adults showed high levels of endorsement for such things as extrasensory perception (41%), haunted houses (37%), ghosts (32%), telepathy (31%), clairvoyance (26%), astrology (25%), witches (21%), and reincarnation (20%) (French and Stone 2014, 6). If we routinely and explicitly used the idea that we should presume nonexistence and that the burden of proof is upon those who make existence claims to show a preponderance of evidence in favor of it, many such beliefs would disappear.

As an example of how this logic can dispel false existence clams, recall the sensational announcement in December 2002 by the group known as the Raelians that they had successfully cloned a human baby whom they had named Eve, and that four other similarly cloned babies were on the way. Dolly the sheep, the first cloned animal, had been born in 1996 and had led to a race to clone more animals, so this technology was much in the news and thus this Raelian announcement was not dismissed out of hand by the media who covered their press conference.

Raelians argue that Darwin's theory of evolution and descent with modification (using the mechanism of random mutation and natural selection) is wrong because life on Earth is too complex to have evolved that way and must have been designed. This same argument is also advanced by theists but for Raelians their designer is not a god. Instead it is a race of extraterrestrials. According to Raelians, on a distant planet there live a highly advanced alien community called the Elohim that long ago had reached an advanced stage of scientific and technical knowledge and developed powerful biological engineering techniques that enabled them to make living cells and to tinker and modify them. They were naturally fearful about letting loose these experimental organisms into their own environment because of the harm they could do, so they looked for a planet that they could use as a laboratory to field test their genetic engineering, to create a home for all their creations so that they could safely see what worked and what didn't. They chose Earth to use as their vast laboratory. They took the then lifeless planet and set about building life on it. Starting with simple cells, they proceeded to create seeds, grasses and other vegetation and progressed to plankton, small fish, then larger fish, then dinosaurs, sea and land creatures, herbivores and carnivores before they tackled the big project, creating beings like themselves. Thus came *homo sapiens*. This, according to Raelians, is how the Earth became populated with all the life forms we see around us (Pennock 1999).

Most people, if they had heard of the Raelian mythology at all, did not take this fanciful scenario seriously but treated it as good, clean, fun. But the claim by the Raelians that they possessed human cloning technology garnered them a blizzard of worldwide publicity and the media rightly asked to see the baby that they claimed proved it. The Raelians said they would produce the baby later, but as time passed and no baby was forthcoming, people reasonably concluded that the whole thing was a hoax. *No one felt obliged to prove that there was no such baby.* The Raelians had made an existence claim and the burden was on them to provide evidence for it. Failing to do so meant that we were perfectly justified in concluding that there was no cloned baby.

So to repeat, in the case of an *existence* claim, the burden of proof is upon the person making the assertion. In the absence of a preponderance of evidence in its favor, the claim can be dismissed. As has often been said, "What can be asserted without evidence can be dismissed without evidence." The basis for this stance is the practical one that, except in very limited circumstances, proving the nonexistence of an entity is impossible. Hence if we do *not* have a preponderance of evidence in favor of the existence of an entity, we can firmly conclude that it is not there.

With universal claims however, the situation is reversed and the burden of proof shifts to the person disputing the universal claim because now *the default position is that the claim is true.* The refutation of a universal claim usually requires the person arguing for the refutation to provide a counterexample that violates the universal claim, along with a preponderance of evidence in support of that example.

Again, this rule serves eminently practical purposes. As an example, consider again the case of the electron. It was initially postulated as an explanation for certain experimental results. In order to probe this, more experiments were done and the properties of the electron (such as its mass and electric charge) were investigated and refined. We have now reached the stage where the existence of the electron and it properties are so firmly established that though we have still not *directly* observed it, to doubt its existence is to lack good sense. But note that in investigating the properties of the electron, the implicit assumption was that *all* electrons have the same properties and behave identically under identical circumstances. But the claim that *all* electrons have identical masses and charges is a *universal* claim and can never be proven to be true with just supporting evidence because we cannot measure the properties of every single electron in the universe. But once the existence of the electron has been established and a few of them act as if they have the same mass and charge, the universal claim that all of them do is *presumed* to be true unless someone comes up with evidence that contradicts it. This is why the proposition "All electrons have the same mass and charge and behave identically in interactions with other particles" is believed to be a true proposition. In science, we believe that natural laws are invariably followed without exception. Universal claims about the properties of an entity whose existence has already been established are taken to be presumptively true unless a preponderance of evidence is provided that contradicts the claim.

It is always within the realm of possibility that someone might come along with data that suggest that there exists a particle that seems to behave identically as the electron but has (say) a different mass. That has actually happened in the past. The scientific community responded to this development with further experimentation that confirmed the existence of this new particle, now called the muon, and it is now considered a true proposition that muons exist and behave similarly to electrons except that their mass differs from that of electrons. The new universal claims are that all muons are identical to each other but different from all electrons.

One cannot back into an existence claim by trying to reframe it as a universal claim, as some try to do. For example, the statement "there is no evidence that a cloned baby does not exist" is a universal statement and thus, by the rules of scientific logic, can be presumed true until there is counterevidence to disprove it. The same is true for the statement advanced by some theists that "you cannot prove that God does not exist." We can all agree that we cannot *disprove* the existence of any entity, especially if believers in its existence reserve the right to ascribe any and all properties to it, including the ability to evade detection. There is no dispute there. The problem is that believers try to use this agreement about that universal statement to then assert that therefore it is *reasonable* to believe that their preferred entity exists. But it is not reasonable to do so because this is asking us to believe in the reasonability of an *existence* statement, and then the burden of proof immediately shifts to advocates to provide a preponderance of positive evidence of existence. As long as they refrain from making that last inference about the reasonableness of existence and stick with the original universal statement that we cannot prove nonexistence, then we can be in agreement.

Here too there is a legal parallel expressed in the Latin maxim *Falsus in uno, falsus in omnibus* that translates as "false in one thing, false in every thing." A witness who takes an oath or affirmation to tell the truth is then presumed to be *always* telling the truth, a universal claim that obviously cannot be proven to be true. A lawyer who wants to challenge the veracity of the witness needs to undermine this universal claim and thus has to show that the witness has lied in at least one important instance relevant to the case. This is an existence claim and the lawyer now bears the burden of proof and has to provide evidence of it.

So to summarize, existence claims can be proven but not disproven, while universal claims can be disproven but not proven, where we must be clear that "proven" simply means that there exists such a preponderance of evidence in support of the claim that we feel safe in assuming it to be true, not that it has been unequivocally proven to be true. But in both cases the rules for establishing existence claims, including the need for a preponderance of evidence in favor of existence, come into play. The reason for these rules about how to judge the truth of existence and universal claims is simply because without them science, and indeed much of everyday life, would be unworkable. In most cases, it is impossible to prove that an existence claim is false or that a universal claim is true and without these rules we would be swamped with claims for the existence of entities and of violations of universal laws with no ability to judge their credibility.

Things are rarely so cut and dry of course. Sometimes one gets into grey areas about what constitutes a counterexample and whether seemingly blanket universal statements contain implicit caveats that limit their generality. For example, consider the universal statement that all cows have four legs. Most people would confidently assert that this is true since they have not seen anything else. But

THE BURDEN OF PROOF IN SCIENTIFIC AND LEGAL SYSTEMS 195

because cows have, albeit rarely, been born with a different number (one even having seven legs), the statement that all cows have four legs is not true, and one has to add caveats. One can try to salvage a universal statement by saying that under normal circumstances cows have four legs but then one can get into disputes about what constitutes "normal."

As an aside, the Raelian creation story is yet another example of how it is always possible to construct alternative theories to explain any given set of facts. In postulating that life on Earth originated elsewhere in the universe and was later introduced here, the Raelians were adopting a highly imaginative, elaborate, and detailed form of the panspermia model of the extraterrestrial origins of life on Earth. It is impossible to refute it because to do so requires knowledge of every corner of the universe. Of course, as with all such panspermia theories, what it does is shift the problem of the origin of life on Earth to the origin of life elsewhere, but that kind of regress is usually ignored by advocates.

The Raelian theory is a comprehensive and naturalistic explanation of how all life evolved on Earth. It can explain *everything* about the living world. But it is a useless theory because it generates no predictions that can be tested. The only basis by which to judge which theory is worth taking seriously is the quality of the evidence produced in support of it, *not on how many facts it purportedly explains*. As long as a preponderance of evidence is lacking about the extraterrestrial origins of life on Earth, we are justified in treating the Raelian theory as false.

That is exactly how things should work and how we should treat arguments. *Logic and reason alone cannot establish the existence of any entity*. One always needs to provide a preponderance of evidence in its favor. As we will see in the next chapter, a purely logical argument can be used to establish the truth of a *proposition* by showing that the *negation* of that proposition leads to a logical contradiction, though this method cannot be establish the existence of any *entity*.

15

Proof by logical contradiction

[This chapter looks at mathematical proofs that use the method of *logical contradiction* and examines how far this can be taken in science. While the existence of any entity can never be proven by this method, the *nonexistence* of certain entities can.]

There is one further avenue to arriving at the truth of propositions that we need to explore because it involves using logic alone and no empirical evidence. The method originates in mathematics and it is useful to contrast its use in science with its use in theology. Theologians often try to claim that they can arrive at eternal truths about the existence or properties of God using pure logic. In some sense, they are forced to make this claim because the empirical evidence at their disposal is either weak or nonexistent. But given my argument in the previous chapter that the burden of providing a preponderance of evidence lies with those making the existence claim, it is worthwhile to examine more closely whether it is possible to prove *purely logically* the existence of *any* entity. If so, we can see if that method can be co-opted into science, thus bypassing the need for evidence in at least some cases.

In mathematics, one way to prove purely logically that a theorem is true is the method known as *reductio ad absurdum* translated as "reduction to absurdity." Suppose we suspect that some proposition is true and want to prove it. Instead of the tedious process of starting from the axioms and proving it as a theorem, we start by assuming that the *negation* of that proposition is true, and then show that this leads to a *logical contradiction or a result that is manifestly false*. This would convincingly prove that our starting assumption (the negation of the proposition under consideration) was false and hence that the original proposition must be true.

A famous example of this kind of proof is the short, simple, and elegant proof of the proposition that the square root of 2 is *not* a rational number, a proof that dates back to Pythagorean times (c. 500 B.C.E.). This proof starts by assuming that the negation of that proposition is true, that is, that the square root of two *is* a rational number. You can then show that this assumption leads to a logical contradiction. Hence it is not a rational number. (I believe that everyone should know this beautiful proof and so have provided it in Supplementary Materials D.)

Another beautiful and simple proof that also uses the *reductio ad absurdum* logic is the one that is ascribed to the mathematician Euclid that there is no limit to the number of prime numbers. A prime number is any number that is 2 or larger that can only be divided by itself or the number 1. Starting from the smallest, they are 2, 3, 5, 7, 11, 13, 17, 19, 23, . . . The question is whether the series of prime numbers ends at some point or increases indefinitely. To prove that it doesn't end, we start with the opposite assumption that it does, and again this leads to a contradiction. Hence the prime numbers must increase indefinitely. (This proof can be found in Supplementary Materials E.)

Using the same logic within the world of science, Galileo and some of his predecessors used a contradiction argument to counter what seemed to be a common-sense belief, that in free fall a heavier object would fall at a faster rate than a lighter one (Butterfield 1957, 82). After all, our everyday experience shows us that leaves and feathers fall more slowly than rocks. The logical argument to prove that *all* objects in free fall must fall at the same rate starts by assuming the opposite to be true, that they would fall at different rates with a heavier object falling faster than a lighter one. We then imagine connecting the heavier and lighter objects with a string to form a third, even heavier, composite object. The slower moving lighter mass should now act as a drag on its heavier partner, making it fall more slowly than if it were falling on its own. But since the composite object has a greater mass than the heavier component, according to our starting assumption it should fall faster than the heavier object falling alone. Hence there is a contradiction that is resolved by assuming that all masses fall at the same rate. The reason why feathers and rocks fall at different rates is ascribed to the effect of air resistance that slows down the feather more than the rock, and experiments that remove this drag factor by being conducted in a vacuum show that they do indeed fall at the same rate.

Note that we have proven a scientific *proposition* to be true without appealing to any experimental data or the "real" world. As far as I am aware, the only way to prove that a *proposition* is true *using pure logic alone* is of this nature, to show that the negation of the proposition leads to a logical contradiction.

Philosophers and theologians down the ages have tried to apply this *reductio ad absurdum* argument to prove the existence of gods using logic alone. But when it comes to *entities*, this method only enables us to establish nonexistence, not existence. The problem, as philosopher David Hume pointed out a long time ago, is that any attempt to demonstrate, as a matter of empirical fact, the existence of *any* entity by a priori arguments such as this will fail because assuming that there is no entity, even God, does not lead to a *logical* contradiction.

There is an evident absurdity in pretending to demonstrate a matter of fact, or to prove it by arguments a priori. *Nothing is demonstrable, unless the contrary is a contradiction.* Nothing, that is directly conceivable, implies a contradiction. *Whatever we conceive as existent, we can also conceive as non-existent.*

> There is no being, therefore, whose non-existence implies a contradiction.
> Consequently there is no Being whose contradiction is demonstrable. (Hume
> 1779, my emphasis)

Given that the nonexistence of God does not lead to a logical contradiction,
theologians resort to the next best thing and appeal to what they feel is manifestly true, that the assumption that God does not exist means that the existence and properties of at least some features of the universe are inexplicable.
This is an argument that has been around for a long time and is expressed in its
most popular form by Paley's watch that we encountered earlier, and has been
resurrected more recently by intelligent design advocates. Almost all supposedly
logical arguments for the existence of God are at some level appeals to this kind
of incredulity.

But the argument that the world is inexplicable without postulating the existence of a designer is not a purely *logical* contradiction, since it is after all appealing to the *empirical* properties of the universe. In days gone by, when much of
how the world works must have seemed deeply mysterious, this subtle equating
of empirical incredulity with logical contradiction may have passed without
much notice. Even if what was shown was not strictly a logical contradiction, the
negation of the proposition "God exists" seemed to suggest that the properties
of the world were wholly or largely inexplicable. Hence such a negation could be
rejected, thus "proving" the original proposition to be true and that God exists.
But this conclusion is not due to a logical contradiction and such arguments
no longer carry any weight since science has explanations for much of how the
world works. *Assuming that God does not exist no longer leads to either a logical
or empirical contradiction.* Establishing the existence of God, like any other existence claim, requires a preponderance of positive evidence in its favor.

Although pure logic cannot be used to establish the existence of any entity,
it can be used to establish some facts of *nonexistence*, since some things cannot
exist because the way they are defined leads to a logical contradiction. For example, we can say that square triangles don't exist. In a similar manner, some
ancient Greek philosophers pointed out that certain properties ascribed to gods
led to contradictions and thus could not all be true. Epicurus (341–271 B.C.E.)
said that the existence of evil was in fundamental contradiction with properties
that are commonly ascribed to gods, such as simultaneously having the qualities
of goodness and omnipotence.

> Is god willing to prevent evil but not able? Then he is not omnipotent.
> Is he able but not willing? Then he is malevolent.
> Is god both able and willing? Then whence cometh evil?
> Is he neither able nor willing? Then why call him god?

In this case, Epicurus is also partially appealing to empirical evidence by asserting the existence of evil as an incontrovertible fact. But his argument can be taken further without resorting to such evidence. As mathematician John Allen Paulos has pointed out, believing that God is both omnipotent and omniscient leads to an immediate contradiction purely on the basis of logic.

> Being omniscient, God knows everything that will happen; He can predict the future trajectory of every snowflake, the sprouting of every blade of grass, and the deeds of every human being, as well as all of His own actions. But being omnipotent, He can act in any way and do anything He wants, including behaving in ways different from those He'd predicted, making his expectations uncertain and fallible. He thus can't be both omnipotent and omniscient. (Paulos 2008, 41)

This contradiction is similar in spirit to the familiar conundrum as to whether an omnipotent God can create a rock that is too heavy for him to lift, because both "yes" and "no" answers contradict the omnipotence property.

As Hume said, anything that can be conceived to exist can also be conceived to *not* exist without any contradiction, so positive empirical evidence must be produced in support of existence. But anything that can be logically excluded, like a four-cornered triangle or a god who is simultaneously omnipotent and omniscient, can be safely assumed to not exist. In science, judgments about which entities exists and which theories and propositions are true and which are false have to be made on the quality of the evidence and arguments that relate "theoretical predictions" with "empirical facts," bearing in mind that each of those things consist, as will be discussed in chapter 18, of *packages* combining a multiplicity of theories, assumptions, and hypotheses.

Religious apologists often resort to what can be called the "mysterious ways clause," that says that certain things cannot be understood using evidence and logic and reason but are instead deep mysteries that are beyond the capacity of mere mortals to comprehend and can only be understood if revealed by God, presumably in the hereafter when we are dead. In the absence of such revelations, one must take them on faith. The idea that certain things that are logical contradictions can exist but are mysteries that cannot be understood except by revelation is not something that can be accepted within the framework set by scientific logic for the reasons given in chapter 10(e). It would be like claiming that (say) four-cornered triangles exist. No scientist or mathematician would suggest that they exist but their existence is a deep mystery that can be understood only by faith. Instead they would be dismissed as human-created self-contradictory entities whose existence is logically excluded.

It is not that the word "mystery" has no place in science. When it comes to those areas of knowledge where we are still largely ignorant, a useful distinction to make is between problems (or puzzles) and mysteries. Linguist Noam Chomsky made the distinction in the context of human language acquisition and the mind, but it is more broadly applicable.

> I would like to distinguish roughly between two kinds of issues that arise in the study of language and mind: those that appear to be within reach of approaches and concepts that are moderately well understood—what I will call "problems"; and others that remain as obscure to us today as when they were originally formulated—what I will call "mysteries." The distinction reflects in part a subjective evaluation of what has been achieved or might be achieved in terms of ideas now available.
>
> ...
>
> [When it comes to problems] much is unknown. In this sense there are many mysteries here. But we have a certain grasp of the problem, and can make progress by posing and sometimes answering questions that arise along the way, with at least some degree of confidence that we know what we are doing. (Chomsky 1975, 137–38)

The transition of something from being a mystery to a problem can be found frequently in all of science but particularly in the field of medicine, where a mystery can arise when a patient comes in with a set of symptoms that is nothing like what physicians have seen before. They will make guesses about what the problem is based on similarities with what they already know, and try out various tests and treatments based on those hypotheses, hoping that something will turn up that points to the source of the problem. But once a firm diagnosis is made as to a possible cause, the case becomes transformed from a mystery to a puzzle or problem. Even though they may not as yet have a solution or treatment, the very fact that it has become a concrete problem enables them to use all the tools and techniques at their disposal to attack it. Much of the history of medicine, the case of HIV/AIDS being a recent example, follows this path from mystery to problem/puzzle to treatment.

When one looks back at the evolution of scientific knowledge, one similarly sees the steady transformation of things that were once mysteries becoming problems, and once they become problems, they usually end up largely being solved and becoming knowledge. Indeed, Francis Bacon (1561–1626) argued in his essay *The Refutation of Philosophies* that it is precisely the role of science to eliminate mysteries, saying "The mark of genuine science is that its explanations take the mystery out of things. Imposture dresses things up to seem more wonderful than they would be without the dress" (Farrington 1964, 123; Ball 2012, 100).

What causes a mystery to transition to becoming a problem is not clear but one marker that such a transition has occurred is with the adoption of a paradigm in that field, because once that happens, the research questions become more sharply posed, tools and expertise for investigating those questions are developed, criteria for judging the quality of proposed solutions become well-defined, and a community of scientists who are deemed competent to make those judgments comes into being.

Scientists abhor mysteries and seek to eliminate them, seeing them as temporary states of ignorance that will be eliminated in due course. The idea that there are some questions that are *permanently* outside the scope of investigation goes deeply against the scientific grain since nothing is considered out of bounds for study. Thus the number of mysteries has steadily decreased over time. The last remaining scientific questions that had some claim to being mysteries are the nature of consciousness and the mind, the origin of life, the origin of the universe, and the source of morality. But in all these areas we see that they are already transitioning to becoming problems. There is a vast research literature on all these four areas and one can find general introductions on recent developments on consciousness and the mind (Blackmore 2005; Dennett 1991, 2017), on the origin of life (Hazen 2005; Pross 2012; Ricardo and Szostak 2009), on the origins of the universe (Hawking and Mlodinow 2010), and on the evolution of the moral sense (Boehm 2012; Krebs 2011; Katz 2000). Scientists no longer throw up their hands and plead total ignorance for any of those areas, as would be the case if they were still mysteries. In each case, there are teams of researchers exploring various theories and testing them out. For the origin of life, for example, scientists have developed an increased understanding of the *process* by which life could have emerged, even though they have not as yet been able to synthesize it or lay out the actual historical process. We may not have definitive answers to all those questions as yet but if history is any guide, once a mystery makes the transition to becoming a problem, it is only a matter of time before we have a plausible solution.

But not everyone is pleased with the steady elimination of mysteries. Some like to believe in mysteries and to even wallow in them and, like Hamlet saying "There are more things in heaven and earth, Horatio, than are dreamt of in your philosophy," think that there are permanent mysteries whose secrets will be forever hidden from us. I was on an academic panel once that was exploring the science–religion issue and I described the process of how in science we have seen in so many areas the steady evolution from mystery to problem and then from problem to solution, and that there was no reason to think that this process would not continue until there were no remaining mysteries, though we likely will never solve all the problems. During the question period, I was asked whether by eliminating all mysteries, science was not also removing a sense of wonder about the world. When we view a beautiful sunset or listen to a piece of exquisite music, I was told, the experience seems transcendental and beyond the power of mere words to describe. But if we could analyze and understand those experiences, would we be also destroying the sense of wonder generated by them? This is an old concern, dating back at least as far as the dawn of the modern scientific era in the seventeenth century when the idea of a mechanical universe, with its workings as potentially understandable as that of a clock, gained popularity and threatened to take the wonder out of our understanding of nature and lead to disenchantment (Shapin 1996, 36).

That is an interesting point of view but I must admit to being puzzled by this idea that ignorance is a necessary condition to view with wonder and awe the world around us. For me, understanding the world in scientific terms, rather than detracting from the sense of wonder, actually adds to it. The fact that we now know why the sky is blue and the sunset is red and that we understand why stars twinkle and how light waves from distant stars travel through vast reaches of space and enter our eyes and then register in our brains *increases* the wonder of it all, rather than decreases it. Similarly, we now understand the basic process of how the sound waves created by an orchestra spread out over space and a part of them enters through our ears that then trigger electrical impulses that go to specific regions of the brain and create the sensation of sound in our heads. Einstein once said, "It would be possible to describe everything scientifically, but it would make no sense; it would be without meaning, as if you described a Beethoven symphony as a variation of wave pressure." But this dual understanding makes it *more* meaningful to me, not less. I personally feel enriched when experiencing phenomena on two different levels, one the purely sensory and the other the intellectual. They complement each other rather than compete.

There seems to me to be little merit in restricting the scope of knowledge and forbidding access to certain areas of knowledge so as to retain mysteries for their own sake. It seems like those who advocate for such limits are fearful of what they might find, that the process may eliminate cherished beliefs. But science knows no such limits and goes where curiosity takes it, breaking down barrier after barrier.

16

The role of negative evidence
in establishing universal claims

[This chapter looks at the important role that *negative* evidence, the things we do *not* observe, plays in scientific logic. This is illustrated by the example of why we so strongly believe that only two kinds of electric charges exist, to the extent of basing our entire modern technology on it, even though we have not proved it to be so, and indeed cannot even hope to do so.]

In certain scientific contexts, *negative* evidence can be powerful in the same way that it can be in the legal setting. A well-known example of negative evidence occurs in a short piece of dialogue in the famous Sherlock Holmes short story "Silver Blaze," where a police officer asks Holmes for guidance.

> "Is there any point to which you would wish to draw my attention?"
> "To the curious incident of the dog in the night-time."
> "The dog did nothing in the night-time."
> "That was the curious incident," remarked Sherlock Holmes.

Holmes draws a strong inference from a dog that did *not* bark in the night. Since our expectation is that dogs bark when unexpected events occur in the night, we can infer from a silent dog that nothing untoward happened. The example of electric charge similarly illustrates the important role that negative evidence, those things that we do *not* see, plays, showing that just because scientific proofs do not have the same status as mathematical proofs does not mean that scientific conclusions cannot be extremely robust.

Most people would readily accept that there are just two kinds of electric charge, positive and negative. This forms the basis of pretty much all of modern technology, because electromagnetic theory is based on it. This is one of the most firmly held beliefs in all of science and almost the entire modern world is constructed on the basis of this two-charge model. Thus it is about as well-established a "fact" as one is likely to find in science and no one even thinks of questioning it. *But we have not proved it to be true.* There is always the theoretical possibility that there exists a third type of charge that we have not as yet discovered. We cannot *empirically* or *logically* rule out that proposition, any more than we can purely logically rule out the possibility of existence of the Loch Ness monster or the Yeti or ghosts and vampires. After all, there is no law that guarantees that only two kinds of charge can exist. The names positive (+) and negative (–) that we give to electric charges seem to rule out any other kinds but those are just labels that turn out to be mathematically convenient. We could have labeled the two kinds of electric charges as red and blue or green and anti-green, and those choices would have allowed us to add more charges if they were discovered. In fact, when it comes to the *strong* interactions between the subnuclear particles known as quarks, there are three kinds of "charges" (not electric) that have been assigned the labels red, green, and blue and have their corresponding opposite charges labeled anti-red, anti-green, and anti-blue.

Do we have absolute certainty that there are only two kinds of electric charge? Have we looked everywhere and convinced ourselves of this? The answer to both these questions is no. Furthermore, we *cannot* prove it to be true because the nonexistence of a third kind of charge is a *universal* claim that cannot be established affirmatively. So how can we be so certain that there are just two kinds of electric charges that we base our entire electricity-based technology on it? We believe it to be the case by using *negative evidence*, because there is an *absence* of the kind of data and evidence that would contradict it.

Here's how that argument works. Suppose you have three electrically charged objects A, B, and C. What scientists find empirically is that *if* the charges are such that A and B attract each other and A and C attract each other, then it is *always* found that B and C repel each other. This set of three observations can be explained by: (1) *postulating* that A, B, and C consist of just two kinds of charges; and (2) *inventing a law* that says that like charges repel each other and unlike charges attract each other. These form the "axioms" of our system. Using these axioms, the reasoning goes that since A and B attract each other, then B must have a charge unlike that of A. Since A and C attract each other, that means that C also has a charge unlike that of A. Since there are only two kinds of charge, that means that B and C must have charges that are alike. Hence B and C must repel. Since all the charges we are aware of follow this pattern, we make the universal claim that only two kinds of charges exist and that they follow this law.

If a set of three charges A, B, and C were ever found such that B and C did not repel each other even though each attracted A, then that would be evidence for the possible existence of a third kind of charge or that there is another rule governing their interactions. In other words, one or both of our axioms would be false. If someone wants to suggest either of those possibilities, then the onus is upon them to come up with credible evidence of it. No one has done so yet.

Because of this *absence* of any evidence that contradicts predictions based on these two axioms about the nature of electric charges, scientists will say that they are extremely confident *to the point of certainty* that there are only two kinds of charges and this is all the "proof" they need. They will proceed on the assumption that this third charge does not exist, and it is perfectly rational for them to do so. But note that haven't actually proved it in any mathematical sense. It is just *a powerful inference* based on the *absence* of certain kinds of data, of dogs that did not bark in the night, if you will. But it is sufficient to convince scientists that the statements are true because *that is how scientific logic works.* We have constructed the entire edifice of electromagnetic theory and much of modern science and technology on that foundation, and the overwhelming success of that enterprise is a testament to the soundness, not just of that particular theory, but of the inferential reasoning used in scientific logic that works backward from comparisons of predictions with data to make judgments about the truth of the existence and nonexistence of entities and the validity of the laws that govern them.

Notice though that the "proof" that only two kinds of charges exist can be challenged by determined skeptics. After all, we have done such experiments with just a few sets of charges. We have not exhaustively repeated them with *every single charge that exists in the universe* because it would be impossible to do so. As a result, someone can say that scientists are wrong, that there does exist a third kind of charge but that it has not been found yet or that it does not interfere with the experiments that scientists do or that its effects are somehow hidden. There is no way that scientists can prove this person wrong. How could they? But what they will do is ignore this argument as not worth responding to because that kind of evidence-free argument has the same standing as claims that a unicorn lives in my home but has the magical ability to remain undetected. Philosophers of science have long pointed out that there is no proposition, however idiotic, that cannot be made immune from refutation by the addition of a protective belt of auxiliary hypotheses that shield its weaknesses. If you want to convince scientists that something like a third kind of charge exists, you will have to provide credible, *positive* evidence, some actual data that cannot be explained by the two-charge theory. *A belief that is not testable and predicts nothing observable is of no use to scientists,* and they will ignore it. A person who clings to a belief in a mysterious third charge that has no observable consequences will be treated as somewhat eccentric.

The importance of negative evidence is most manifest when it comes to the *laws* of science. These are always phrased as universal claims because they assert that they will be followed at all times and under all conditions. It is the absence of any evidence to the contrary that is used to justify them. If someone wants to refute that claim of universality, the burden of proof is upon them to provide credible evidence of a counterexample that violates the law. If convincing evidence for such an exception is found, the law is revised to allow for the specific conditions under which it does not hold, but even those exceptions are not idiosyncratic and the law is still phrased in universal terms.

For example, there exist a number of what are called "conservation laws" in physics. What a conservation law asserts is that there is a property whose *total* value for a closed system stays the same (i.e., is "conserved") even as other things change. *Conservation laws are universal statements and are believed to be true purely because no violations of them have ever been observed.* We are most familiar with energy conservation, although there are many others. The law of energy conservation states that for a closed system, energy can change from one type to another but the total value will remain unchanged. A simple illustration is where we run an electric motor using a battery. In that system, chemical energy stored in the battery is transformed to electrical energy that is then converted to mechanical energy when the battery runs the motor, and the mechanical energy becomes heat when the running motor gets hot. But if we add up all the different forms of energy in the system at any one point in time, the value of that sum will remain the same throughout the process, even as the amount in each form of it changes from instant to instant.

The existence of a universal law is sometimes suggested by a rule of thumb that says that "if something *can* happen, it *will* happen." The corollary to this is that if we never observe something that we might expect to see, then *there must be a law that prevents it*, another example of the use of negative evidence. Hence if some thing that could happen is never observed, its absence is explained by postulating the existence of a law that such an occurrence would violate. Many of the conservation laws (such as those for electric charge) originated in this manner, and subsequent theories include those laws as part of their structure. Anyone who challenges the validity of these conservation laws bears the burden of proof to provide evidence of such a violation. This approach to establishing the validity of conservation laws by the absence of contradictory evidence is so routinely followed in science that few even bother to state it explicitly. (Supplementary Materials F looks at the important role that conservation laws play in elementary particle physics.)

So in science this interplay of existence and universal claims, and the different ways they are established including the use of negative evidence, goes on all the time and forms an integral part of scientific logic and the way that scientific knowledge is constructed. Being aware of this kind of reasoning is invaluable in reaching sound conclusions in *all areas of life for everyone*, which is why increasing general awareness of them is so important. Scientific ways of thinking benefit everyone.

It necessarily follows from this discussion that in science we use the word "true" provisionally and not absolutely. In the case of existence claims, "true" is taken to mean that existence is supported by a preponderance of evidence. In the case of universal claims, "true" is used as shorthand for "not as yet shown by a preponderance of evidence to be contradicted." In the next chapter, we will look at how the logic of science is used to make inferences in four areas currently at the frontier of physics research: dark matter, dark energy, string theory, and the multiverse.

17

Dark matter, dark energy, string theory, and the multiverse

[This chapter uses what we have learned about scientific logic to evaluate the status of four theories that are currently at the frontiers of physics and command much attention in the media: dark matter, dark energy, string theory, and the multiverse.]

At this point it is perhaps appropriate to discuss the status of dark matter, dark energy, string theory, and the multiverse, four major areas of physics research that have received much attention because of the counterintuitive, even bizarre, nature of the existence and universal claims that have been made but where there have been as yet no definitive conclusions as to their validity. These four cases show how science proceeds at the frontiers of knowledge, with the logic of science enabling researchers to make inferences and tentative hypotheses that they can test and thus guide their investigations. I will separate these four areas into two groups, dark matter and dark energy on the one hand, and string theory and the multiverse on the other, because the reasons and evidence for each group are of different kinds and strengths.

In the case of the first pair of dark matter and dark energy, it is currently believed that "normal" matter and energy, that is, those that we are familiar with such as atoms and molecules and the more elementary particles that comprise them such as protons, neutrons, electrons, and all the other particles whose existence has been established, constitute a surprisingly small percentage of all the mass and energy that make up the universe, just about 5% in fact. Dark matter is a new kind of matter and is thought to make up about 26% of the mass-energy of the universe. Dark energy is a new form of energy that is similarly postulated to exist and makes up the remaining 69% of the universe. Neither dark matter nor dark energy has been positively detected and their existence is only inferred. Why they have been given provisional existence status illustrates the use of scientific logic.

The idea of dark matter originated with the detection of anomalies in the velocities of stars in galaxies, with them having values that differed from predictions based on the observed amount of familiar matter that we knew about. The proposed solution was that there was a significant amount of matter in the universe that we had *not* detected and accounted for, and that it was this new matter that caused this anomalous behavior. This idea had been around from the first quarter of the twentieth century but received a major boost in the 1970s with more careful measurements using new technology by Vera Rubin and coworkers of the speeds of stars in the spiral arms of galaxies (Rubin and Ford 1970). These more precise data could not be explained using existing, well-established theories of gravity using the "visible" (i.e., known) amounts of matter but could be explained by postulating that galaxies were immersed in a sphere of matter that we had not detected as yet that extends well beyond the visible edge of the galaxies. The mass of this dark matter has been postulated as the source of the gravitational force needed to produce this anomalous behavior. Subsequent measures of stellar phenomena have bolstered the idea of such matter existing.

So if this matter is so abundant and we are immersed in it, why have we not directly felt or observed it? This is indeed a pressing question in the field of astrophysics and cosmology and a multiplicity of theories have emerged that seek to explain the amount of dark matter, what it might consist of, why we have not observed it so far, and what possible experiments might be performed to directly detect these dark matter particles. Clearly, if they exist, their interactions with ordinary matter must be extremely weak since it is all around us and yet we do not feel it. Hence the major searches currently underway involve setting up heavily shielded, highly sensitive detectors in deep underground mines to minimize the possibility of other cosmic particles falsely triggering the detectors.

The current status of dark matter is similar to that of the early days of neutrinos. The existence of those earlier particles had been postulated to explain why some observed nuclear reactions seemed to violate the laws of conservation of energy and angular momentum. Arguments for the existence of that hitherto unknown particle competed with theories that suggested that the conservation laws needed to be modified for these reactions. Over time, the indirect evidence for the existence of neutrinos was strong enough that, combined with reluctance to abandon the conservation laws, they were given *provisional* status of existence. But the neutrino theory did not merely explain existing anomalies. It also predicted, as all good scientific theories do, results of specific reactions that experimenters could investigate and the testing of these predictions eventually led a quarter century later to direct *positive* evidence of their existence, shifting their status from provisional to confirmed.

Dark matter has similarly become an *explanatory concept* for various stellar phenomena involving galaxies and clusters of galaxies that seem otherwise inexplicable. Some observations of the cosmic microwave background radiation also seem to support the idea of dark matter (Ade et al. 2016). The case for its existence has become sufficiently strong that although we do not know many of its specific features, it has been granted provisional existence status pending direct observation.

Dark energy has similarities with dark matter but also important differences. The similarity is that its existence is also *inferred* from cosmic observations. In this case the surprising finding that distant galaxies are *accelerating* away from us (i.e., speeding up) is a puzzle to which dark energy was proposed as a solution. Dark energy is believed to be a form of energy that occupies all of space but, unlike the other energy types that we are familiar with, has the unusual property that it does not get diluted even as the space it occupies expands. This constant energy density provides the "force" that drives the acceleration of these distant galaxies. As with dark matter, some observations of the cosmic microwave background radiation also seem to support the idea of dark energy (Ade et al. 2016).

While dark matter is thought to be restricted to spherical regions encompassing galaxies, dark energy is thought to occupy all of space in the entire universe. Unlike dark matter particles that are believed to have mass, dark energy is massless. In those two respects of being massless and occupying all space, dark energy is similar to the old aether and, as was done in the nineteenth century, it has been given *provisional* existence status despite not having been directly detected, because it is seen by many to be necessary as an explanatory concept for certain astrophysical phenomena.

Both dark matter and dark energy make their presence felt via their gravitational effects. They are inferences arrived at using existing theories such as Newton's theory of gravity and Einstein's general theory of relativity, and have been introduced as auxiliary hypotheses to explain anomalous behavior within the framework of those older theories. It is important to emphasize that there is a long chain of inferential reasoning involved in arriving at conclusions about the existence of both dark matter and dark energy.

While the direct detection of either dark matter or dark energy will swing the argument for existence decisively in its favor, the failure of current searches would not imply that they do not exist because of the eternal problem of establishing a negative. Like the aether, they will only cease to "exist" if, in addition to not being directly detected, alternative theories come along that *make them unnecessary as explanatory concepts* (Singham 2011), and these alternative theories also develop a preponderance of positive evidence in their favor. This is how, as was discussed earlier, the theories of special and general relativity eliminated the aether and the planet Vulcan, and the oxygen theory of combustion got rid of phlogiston.

No amount of data can uniquely determine a theory. One can always construct alternative theories to explain any set of facts and that is the case here too. There are competing theories that do not postulate the existence of dark matter or dark energy to explain some of the phenomena. One, known as Modification of Newtonian Dynamics (MOND), claims that slightly modifying the force of Newtonian gravity allows one to explain much of the same phenomena that dark matter does (Milgrom 2002; Milgrom 2014). The recent failure of the most sensitive experiments to date for the direct detection of dark matter, the LUX collaboration experiment in the United States (Akerib et al. 2017) and the PandaX-II Collaboration in China (Tan et al. 2016), have provided hope for those favoring the MOND alternative (Kroupa 2016). Another theory claims to explain the acceleration of distant galaxies without invoking dark energy (Kolb et al. 2006).

In both cases we are confronted with having to choose between keeping current laws and postulating new entities or postulating a change in the laws that govern known entities. This kind of competition for acceptance is not unusual in situations that are still in flux. Alternative theories to dark matter and dark energy have not gained widespread acceptance as yet, largely because they fail to explain many other features of observational data. But that too is not unusual with new theories. It takes time for them, if they have merit, to attract new researchers who will flesh them out by producing results that make them more comprehensive. Since we are never 100% certain of our theories in science, alternative theories will always be around as long as there is no conclusive evidence in favor of dominant theories like those of dark matter and dark energy, especially in the form of direct detection of its entities. The level of interest in alternative theories depends on the success, of lack of it, of the investigations seeking evidence in support of the dominant theory.

String and multiverse theories have a different status from those of dark matter and dark energy. These are indeed new theories that are not the result of trying to explain anomalies in the data but consist rather of attempts to address deeper questions such as the origins of the universe and why our universe is the way it is. Why is there something rather than nothing? Why do we seem to have only three dimensions of space? Why does time seem to flow in only one direction? How can we unify the theory of gravity with quantum mechanics, a major unsolved problem in physics? Why are there only three sets (called "families") of elementary particles? Why do we need at least as many as twenty-six independent fundamental constants whose values have to be experimentally determined? Why are those fundamental properties seemingly fine-tuned to enable the eventual emergence of a universe like ours from the Big Bang? And so on.

String theory suggests that all matter is made up, at the most fundamental level, of tiny, vibrating, interacting strings, and its advocates suggest that it could make possible the long-sought goal of unifying gravity with quantum mechanics. The multiverse theory, the most popular version of which goes by the name of M-theory that actually represents a family of theories (Hawking and Mlodinow 2010, 8), suggests that our universe is not unique and that new universes are constantly being born from existing ones and that there may be as many as 10^{500} of them, a staggeringly large number, with each of them governed by their own set of laws (Hawking and Mlodinow 2010, 118).

While the questions these two theories seek to answer are deep and important and undoubtedly interesting, these theories are at best suggestive and extremely speculative at this point and the predictions that emerge from them by which they might be tested are currently far from feasibility. Critics of those theories sometimes claim that they are so far away from the possibility of direct testing that perhaps they should not be considered science at all (Smolin 2006, 168–70). But indirect tests do exist. For example, string theory suggests that a new set of particles called supersymmetric particles should be produced at the Large Hadron Collider at around the energies that produced the Higgs particle. But those new particles have not been seen as yet, putting a damper on the enthusiasm for string theory but giving a boost to the hopes of alternative theories that do not predict the existence of such particles. One alternative theory is known as "loop quantum gravity" (Rovelli 2016, 214–17). The important point for the purposes of this book is that since scientific theories can never be proven to be true and are always underdetermined by data, *they always have understudies waiting in the wings*, waiting to take center stage if the star should falter.

Furthermore, even the need for string and multiverse theories has been questioned. It may be that the answers to those deeper questions that they are designed to address, such as the fundamental properties of space, time, matter, and origins of the universe, might be just brute facts of nature without any underlying reasons or simpler structure. Since the number of possible universes that might have been created is such a huge number, it is quite possible that one of them, purely by chance, threw up values of the fundamental quantities that enabled a universe with its life to appear. Each universe had to have *some* properties when it came into being and those of our universe just happened to be the ones that produced the kind of life that eventually enabled sentient beings to emerge who are able to ask these kinds of questions. There need be no deeper explanation for our existence.

While the search for deeper and simpler structures have been fruitful avenues of research in the past and have led to all manner of important discoveries, such as DNA and RNA for biology, and the discovery of subatomic and subnuclear particles in physics, those earlier searches were driven by observations that suggested that something smaller did exist. But the problem of induction warns us that just because the search for simpler structures was so successful in the past is no guarantee that it will always be fruitful. So while a good case can be made for the granting provisional existence status to dark matter and dark energy because of the need to explain certain anomalous observations, the case is not as strong for multiverse and string theories, and the attempts by practitioners in those two areas to find answers to deep questions may turn out to be like the search for the chimera of Greek mythology, seeking something that does not exist.

How scientists decide whether to continue the search for evidence within the framework of an existing theory or whether to abandon it in favor of a potentially more fruitful avenue of research predicated on a new theory is a key question that represents a crucial stage in addressing the Great Paradox of science and the last part of the book will discuss how they go about it.

PART FIVE

RESOLVING THE GREAT PARADOX

18

How scientists choose
between competing theories

[Part Five of the book begins the process of weaving together all the earlier threads, with this chapter looking at what historians, philosophers, and sociologists of science have uncovered about the way that the scientific community chooses between competing theories and how the process proceeds rationally and systematically even if questions of truth are not determinative.]

We have seen that theories are fundamental to our understanding of science because it is those that have enabled us to achieve so much success. This has led to the central paradox that this book seeks to address: If we cannot *prove* the theories of science to be true or false but are choosing between competing theories using judgments, how can we be so confident of so many of them to the extent that we place our lives at risk if they are false? The answer is that it is the use of *scientific logic* coupled with a *preponderance of evidence* that enables scientists to converge and arrive at *consensus judgments* on what they *believe* to be true. The benefit of such a consensus is best illustrated by a significant difference between science and religion that explains why scientists can eventually resolve disagreements amongst themselves as to which theory should be considered true, while sophisticated theologians, even after millennia, cannot agree as to which religious theory among all those promoted by differing religions is the true one.

There is no question that believing in the existence of gods satisfies the emotional needs of some people. But in the current age, and especially for sophisticated believers, it is not enough to say that one believes in a god merely because it is comforting or fills a vacuum in one's soul. People feel that they need to justify their beliefs in ways that would pass muster with modern science, at least minimally, and so they try to find more acceptable reasons based on logic and reason and empirical evidence. But their attempts come nowhere close to resembling science.

When there are competing scientific theories within a field, a fairly standard process unfolds. Scientists working in that area meet at conferences to share the latest information, isolate the differences between theories, identify tests that might shed light as to which theory is to be preferred, carry out such tests, and then share the results. This will usually result in a new round of such activities with the questions under investigation becoming more refined and pointed. After some time the weight of evidence is such that the proponents of one side either implicitly concede that their theory should be rejected by ceasing to work on them or are pushed to the margins and fail to attract new researchers, resulting in a consensus mainstream verdict in favor of one theory and the slow disappearance of the others. It must be borne in mind that not all scientists act in the way that is described in this chapter, but that these are broad generalizations.

In the case of religion, however, it is significant that even though there are different religions that have starkly different belief systems (which can be considered to be theories about gods), and these have existed for millennia, there has been no such convergence. While it is true that even in science convergence has sometimes taken centuries to emerge, what is significant is that in the case of religion, there is not even the *attempt* to convene conferences to try and arrive at which religious theory is the right one. This is because since evidence plays little or no role in religion, and reason and logic are invoked only when they support one's own case but are discarded in favor of faith when they don't, there is no basis for arriving at that kind of agreement. Unlike in the case of religion, it would be unthinkable for a scientist to argue in favor of her theory by denying the evidence and logic in support of competing theories and telling people that they must have faith in her theory.

The field of theology does not seek to produce tests that can provide positive evidence for the existence of gods or to enable one to choose as to which religion is true. It instead focuses on explaining why there is no such evidence. Religious believers have the freedom to do this because *pure logic is never sufficient to eliminate a theory*. Those who claim that since neither logical nor evidentiary arguments can *disprove* the existence of gods, it is logically *possible* that gods exist, are saying something that cannot be shown to be wrong but would be meaningless in science. In science, *to believe in the existence of an entity requires credible affirmative evidence in favor of it, not the lack of evidence proving nonexistence.*

Science can come to a consensus despite the fact that not all individual scientists on the losing side change their minds. Some of them can be as dogmatic as the most fervent theologian in holding on to their beliefs and as inventive in finding new reasons for doing so, though they will never resort to appealing to supernatural forces or faith. The key difference is that over time, the advocates of a failing scientific theory become less influential, more marginalized, and eventually die out. The next generation of research students tends to avoid signing on to failing theories, so that those theories eventually fade from the scene, to be found only in historical archives.

In her book *Quantum Dialogue*, Mara Beller (1999) examines closely how this process played out in convincing generations of physicists that the Copenhagen interpretation of quantum mechanics was the only viable one, even though others existed then and continue to exist now (Bohm 1957, 109–16; Bell 1987). She examined the rhetorical techniques used by Niels Bohr, Werner Heisenberg, Max Born, and Wolfgang Pauli to marginalize alternatives that were advocated by such eminent scientists as Louis de Broglie, Erwin Schrödinger, Albert Einstein, David Bohm, and later by John Bell. Bohr, like Newton and Ampere before him, relentlessly argued for the *inevitability* of his point of view, that it was forced upon us both by data and by proofs, despite the fact that those claims were not sustainable.

> Because an argument of "finality," or inevitability, is perhaps the strongest way to advance any sort of claim, the rhetoric of finality and inevitability permeates many philosophical arguments in the Copenhagen framework.
>
> ...
>
> One would expect that the proponents of the Copenhagen interpretation were in possession of some very strong arguments, if not for inevitability, at least for high plausibility. Yet a critical reading reveals that all the far-reaching, or one might say far-fetched, claims of inevitability are built on shaky circular arguments, on intuitively appealing but incorrect statements, on metaphorical allusions to quantum "inseparability" and "indivisibility," which have nothing to do with quantum entanglement and nonlocality. (Beller 1999, 196)

The Statistical Interpretation is one such alternative framework that is consistent with all verifiable predictions of quantum theory (Ballentine 1970). Another is the de Broglie-Bohm pilot wave theory (Bohm 1957). A third is known as the transactional interpretation (Cramer 1986) and there are many more (Ball 2018). So how did the Copenhagen interpretation become so dominant that few physicists nowadays are even aware that alternatives exist that have not been refuted? Beller says that all the four advocates of the Copenhagen school also had *institutional power* in that they headed prominent centers of research through which passed many young students and researchers from all over the world. These people were proselytized into this view and then went out into the world to propagate it (Beller 1999, 38–39). The undoubted empirical success of quantum mechanics was used as evidence of its rightness. Even though the other interpretations could also claim successes, they did not attract sufficient supporters to become viable. Hence the Copenhagen interpretation came to be accepted as the only one worth learning and teaching.

This process of evolution of theories leading to convergence is assisted by the fact that in science, unlike in religion, there is no sacred text that must be adhered to and serves to perpetuate one point of view. As much as scientists admire the works of Isaac Newton and Albert Einstein and Charles Darwin, they do not treat their writings as infallible or the final word. Science has moved on since those classic works were written and the original theories have been greatly modified, expanded, and elaborated upon, even if they still bear the names of their originators. Every generation of students is taught the currently articulated version of accepted theories, not the original ones. During their education, few budding researchers are even exposed to the original writings of these scientists. Instead they get their basic knowledge largely from textbooks that represent the *current* consensus among scientists about what these theories mean and how they should be used, and the range of views presented can be narrow and quite far removed from their origins. For example, the noted evolutionary biologist Carl Woese, coproposer in 1977 of the now mainstream but then radical idea that the overall phylogenetic structure of the living world is tripartite, is said to have not read Darwin's most famous book until the year 2000, when Woese was seventy-two years old (Pace, Sapp, and Goldenfeld 2012). As was pointed out in chapter 12, what seems like narrowed vision due to deliberate blinkering in science education can serve a valuable pedagogical purpose and still lead to periodic overthrowing of dominant paradigms.

In the case of religions, however, the ancient and unchanging texts remain front and center, the starting point and focus of every new generation of believers. Furthermore, children are not encouraged to choose their religious beliefs when they are of more mature age, the way that research scientists choose which theories they want to learn about and work with. Religions indoctrinate the next generation of impressionable children with those ancient beliefs when they are very young, thus ensuring that those beliefs persist. Furthermore, there is a vast industry (churches, temples, mosques, priests, theologians, etc.) whose very livelihood depends on those ancient religious ideas being perpetuated. There is no corresponding institutional structure dedicated to perpetuating old scientific theories. Scientists can shift their allegiance from one theory to another without serious risk of losing their jobs while a theologian or priest mostly cannot. Can one imagine the Pope of the Roman Catholic church, or even a parish priest, saying that after some thought he has come to the conclusion that (say) Buddhism is the true religion, and still remaining in that position? Theologians are able to cling to their old ideas and never have to concede that they are wrong, because in the absence of any empirical constraints, their theories are infinitely malleable and they can invoke more and more ad hoc hypotheses to justify their beliefs.

The evolution of theories over time provides the most striking evidence of how differently science and religion operate. If one were able to turn the clock back (say) five hundred years, the world of science that a present-day scientist would encounter would be so different as to be unrecognizable. Even the word "scientist" would not exist. But a theologian who returned would feel right at home and find the situation pretty much unchanged. If asked to roll back all the knowledge gained in the last five hundred years in just one specific field, which would you choose: to erase what we have learned from religion/theology or what we have learned from science? The answer is obvious. It is the willingness of the community of scientists to shift its allegiance to new theories and declare old ones to be wrong that makes science seem to progress and work so well while religions stay largely unchanged.

As was argued in earlier chapters, in science we cannot *prove* the nonexistence of such entities as phlogiston or the aether nor can we *disprove* the geocentric model of the solar system using logic or evidence. So how are scientists able to say with such confidence that current theories (like relativity and gravity) are true and that others of the past (like the aether or phlogiston) are false? As discussed in chapter 7, Pierre Duhem said that we have to appeal to the collective "good sense" of the *scientific community as a whole* to arrive at a judgment as to which theory is better, that it is the community of professionals working in a given scientific area that is the best judge of how to weigh the evidence and decide whether a theory is right or wrong, true or false. The community as a whole is a better judge than any individual member of that community, since scientists are like other people and prone to personal failings that can cloud their judgment, even if they exercise great vigilance over themselves (Duhem 1906).

Duhem's idea that we defer to the scientific community to tell us what is true and what is false may be disturbing in two ways. It makes small groups of expert scientists in narrow fields the arbiters of what constitutes reliable knowledge. It also seems to diminish the status of scientific "truths," reducing them from being objectively determined by data to the products of collective judgment, however competent the community making those judgments may be. Surely there must be more to it than that? After all, science has achieved amazing things. Our entire modern lives are made possible because of the power of scientific theories that form the foundation of technology. In short, science works and works exceedingly well. How can science so effectively bend nature to its will?

One philosopher of science, James Ladyman, argues in favor of the realist position that such success would be inexplicable if the scientific theories we have developed were not true:

> It is the success of science in predicting new and surprising phenomena, and in the application to technology that realists argue would be miraculous if the theories were not, in general, correctly identifying the unobservable entities and processes that underlie what we observe. (Ladyman 2002, 213)

This sentiment that attributes success in the control of our environment to having correct knowledge is not new, dating at least as far back as the seventeenth century and expressed in statements such as Francis Bacon's statement that "Human knowledge and human power meet in one" (Bacon 1620, Aphorisms III) that is abbreviated in the modern day as "knowledge is power." As Steven Shapin says in his book on what is known as the scientific revolution, "the ability of natural philosophical knowledge to yield practical outcomes and to produce the means for technological control were taken as reliable tests of its truth" (Shapin 1996, 130). Such feelings are so strong that people continue to try to find ways to show that scientific theories, if not absolutely true now, must be at least *progressing* along the road to truth. To better understand these more recent efforts, in the next paragraph I will first briefly review what led up to them (which were discussed in more detail in chapter 10).

Early attempts to prove theories true failed. So did efforts to assign numerical probabilities to the likelihood of theories being true in order to more objectively determine which theories were better. Popper's idea of falsification suggested that we could systematically discard wrong theories in favor of newer ones that had the *potential* to be true, even if we could not prove that they were. Popper's model of falsification makes the scientific enterprise seem rational and logical. It also implies that science is progressing along the path to truth by successively eliminating false theories. As such, it appeals to scientists and it should not be surprising that many, if not most, have embraced it. But falsification fails because we can never compare a prediction based on a single theory to a piece of purely empirical data. Falsification also does not describe the actual historical progress of science.

Thomas Kuhn's model said that scientists are usually not looking to test or overthrow an existing paradigm but instead seek to resolve any anomalies they encounter. If a few key ones turn out to be highly resistant, science can enter a period of crisis whereupon individual scientists start switching allegiance to a new paradigm that seems more promising. Kuhn suggests that the switch from seeing the old paradigm as true to seeing it as false and needing to be replaced by the new one is similar to a *gestalt* switch, a sudden realization that is not driven purely by reason and logic.

Kuhn's views aroused considerable passions and even outright hostility, because it seemed to suggest that the way scientists switched allegiance from an old theory to a new was subjective. Some anti-science people (religious and nonreligious alike) have seized on Kuhn's idea that scientific revolutions are not driven purely by objective factors to extend his views well beyond what he envisaged, and claim that science is an inherently irrational activity and that scientific knowledge is just another form of opinion and has no claim to privileged status. Kuhn spent a considerable part of the rest of his life arguing that this was a distortion of his views and that scientific knowledge had justifiable claims to being more reliable because of the ways that science operates (Marcum 2018).

While the general features of scientific revolutions as outlined by Kuhn hold up well, other historians and philosophers have challenged two aspects of his approach: whether a state of crisis is a prerequisite for a new theory to emerge as the new paradigm, and whether the switch to seeing the world in a new way using a new paradigm is as subjective as his analogy of a gestalt switch implies. Imre Lakatos was one of those concerned that Kuhn's model implied a certain amount of irrationality in the way that scientists choose a new paradigm over the old or the way they pick problems to work on. In his major work *Falsification and the Methodology of Scientific Research Programmes* (Lakatos 1970), he argued that scientists *are* rational in the way they choose paradigms, and he proposed a new model (which he called *methodological falsificationism*) that he claimed solved some of the difficulties of Popper's older model (which he called *naïve falsificationism*).

In the naïve falsification model, when there is disagreement between the predictions of a theory and observations or experiment, the theory must be abandoned. Kuhn and Lakatos both agree with Duhem that when such a disagreement occurs, it is not obvious where to place the blame for the failure, so summarily discarding the paradigmatic theory is unwarranted. But Lakatos says that Kuhn places too much reliance on vague psychological processes to explain scientific revolutions and that the process is more rational, that scientists proceed in a *systematic* way in choosing between competing theories. Lakatos agrees with Kuhn that paradigm choice is never simply the result of a contest between theory and experiment. As pointed out in chapter 10(c), at the very least they are *three-cornered fights* between an old paradigm, a new emerging rival, and experimental data. But he disagrees with Kuhn that a crisis within the old paradigm, that is, a widespread recognition that some anomaly is so serious as to cast doubt on the existing paradigm, is a necessary condition for scientists to switch their allegiance to a new one (Lakatos 1970, 9). He argues that even in the absence of a crisis, a new theory can be *rationally* chosen over its predecessor if it (1) *predicts novel facts* that the old theory would have forbidden or would not even have considered; (2) explains *all* the unrefuted content of the old theory within the limits of observational error; and (3) some of its novel predictions are corroborated (Lakatos 1970, 32).

In his zeal to rescue science from the charge of irrationality, Lakatos went too far. Critics have pointed out that the second condition is almost never met since a new paradigm almost always initially explains just a very few facts and ignores or contradicts many of the results of the old one it seeks to replace. Adopting a new paradigm necessarily entails some of the old problems losing their solved status, at least in the short term. It takes time for a new paradigm to build up a large body of evidence in its favor and once that happens, those results of the old paradigm that it does not agree with are either ignored or treated as irrelevant, thereby giving the *illusion* that the new paradigm explains all the old results. For example, nobody cares anymore about old measurements on the elasticity of the aether or the mass of phlogiston.

In Lakatos's model of methodological falsificationism, he reinforces the idea that experimental data is never free of theory. An experimental result in its raw form is simply a sensory observation, such as a dot on a screen, a pointer reading on a meter, a click of a Geiger counter, a track in a bubble chamber, a piece of bone, and so on, none of which have any obvious meaning in and of themselves. In order to give them some meaning, we have to use theories that interpret the raw sensory experience. For example, a fossil bone that is found is useless unless one can determine, among other things, what animal it belongs to and how old it is. Figuring out these things involve invoking a whole host of other theories from diverse fields. In addition, we have to assume that our knowledge about the other elements surrounding the raw data is unproblematic. So empirical results that we call experimental or observational data consist of a *package* that combines *three* components: raw data, interpretive theories, and unproblematic background knowledge.

Meanwhile, a theoretical prediction is never the product of a single theory but also consists of a another package made up of *four* components: the basic theory or theories being investigated (what Lakatos calls the "hard core"), the initial conditions, various auxiliary hypotheses that are needed to actually implement the theory (we can rarely use a theory in its pure form but usually need to make some approximations to arrive at any conclusions), plus the invocation of *ceteris paribus* clauses (roughly meaning "all other things being equal") that cover the possibility that something else we have not considered may be playing a role unbeknownst to us. For example, to understand the origins of the solar system, we need Newton's laws (the basic hard core theory) but we also need to make assumptions about the initial state of the gas from which it was formed (the initial conditions), assume that all the other laws and theories we use in our calculations are correct and have not changed since the time the Earth was formed (auxiliary hypotheses), and that no other unknown factors played a significant role in the formation (the *ceteris paribus* clauses).

Lakatos said that when we say that there is a disagreement between a "theoretical" prediction and "experimental" data, what we are really saying is that that there is a disagreement between the entire package that goes into making the theoretical prediction and the entire package that goes into giving meaning to the raw data, and it is not immediately obvious where in that complex the fault lies. He said that far from choosing between theories arbitrarily or irrationally, scientists have developed ways to *systematically* investigate such disagreements by exploring all the elements contained in each package until they arrive at a satisfactory resolution of the problem, and that this systematic search process is what makes theory choice in science so rational.

He said that scientists use both a *negative heuristic* and a *positive heuristic* to isolate the cause of the disagreement. The negative heuristic says that one must deflect attention *away* from the basic hard core theory when there is an inconsistency between predictions and experiment. In other words, scientists initially look for the culprit amongst all the factors *other than the basic theory*. The positive heuristic consists of a partially articulated set of guidelines that tell scientists how to *systematically* investigate the initial conditions, auxiliary hypotheses, *ceteris paribus* clauses, unproblematic background knowledge, interpretive theories, and even the raw data. Lakatos says that this positive heuristic "saves the scientist from becoming confused by the ocean of anomalies" that always surrounds them (Lakatos 1970, 50). In short, while a few scientists may explore alternative new theories to explain the anomaly, such theories will not be considered for acceptance until *factors other than the basic theory are carefully and systematically probed first* to see if the problem lies somewhere in there. These two heuristic strategies protect the basic theory from being too easily overthrown. This is important because good theories are hard to come by and one must not discard them hastily. It should only be done as a last resort.

Lakatos claimed that this process explains much of scientific activity and *rationally* determines how scientists select problems to work on and how they resolve paradigm conflicts, contrasting it with Kuhn's suggestion that scientists *intuitively* know what to do in such situations. Lakatos is effectively fleshing out both Duhem's "good sense" and the rules of operation by the scientific community that Kuhn alludes to but does not elaborate on, and making it more representative of actual scientific practice.

A good historical example is how the phlogiston theory of combustion, the first dominant paradigm in chemistry, was replaced by the oxygen theory, a transition so important that it is now referred to as the chemical revolution. Supporters of the phlogiston theory were able to construct one ingenious auxiliary assumption after another to explain away those experimental results that seemed to contradict their theory that when an object burned, it released the phlogiston that had been contained within it. It took a long time for those hypotheses to be investigated and rejected (the positive heuristic at work) before the oxygen theory took hold (Butterfield 1957, 195–209; Kitcher 1993).

A more recent example of Lakatos's model at work occurred in September 2011 when there were published claims by reputable scientific research teams to have discovered neutrinos that traveled faster than the speed of light. If true, this would have violated a fundamental tenet of modern physics ever since Einstein proposed his theory of special relativity in 1905, that no object can exceed the speed of light. This limitation is so well known that even many nonphysicists are aware of it. The report of faster-than-light neutrinos created such a sensation that when I attended a university committee meeting that day, colleagues from other departments eagerly questioned me as to what I thought of this development and whether the theory of relativity had indeed been falsified. I replied that I thought that the report was unlikely to be true because the theory of relativity was so well established, but that I would reserve final judgment until more exhaustive tests had been done and the results replicated by other groups. I suspect most scientists respond the same way when a supposed revolutionary discovery is suddenly reported.

There followed a flurry of negative and positive heuristic activity in all the ways that Lakatos described, with some groups looking more closely at the experiment to see if the result was real and replicable, others making the assumption that the result was real and seeking to preserve Einstein's theory by looking to see if there were other factors at play that might explain the surprising result, while a few explored new theories just in case those other investigations turned up nothing and the theory of relativity was in real trouble. And sure enough, about six months later it was discovered that something as mundane as a loose fiber optic cable connection had created the discrepancy in the data that led to the false conclusion. In addition, other groups had tried to replicate the faster-than-light neutrinos and failed. And so the claim of faster-than-light neutrinos was retracted (Cho 2012). It should be emphasized that the way this process played itself out *is the norm and goes on all the time in science*, but usually not in the bright glare of the news media.

Lakatos argues that as long as a basic theory is fruitful, and the positive heuristic provides plenty of avenues for people to investigate and thus steadily produce new facts that both advance knowledge and are useful (a healthy state of affairs that he labels as *progressive*), the basic theory will be retained even if discrepancies persist. This is why Newtonian physics, one of the most fruitful theories of physics of all time, is still with us even though it would be considered falsified using Popper's naive falsification criteria. It is only when a theory runs out of steam and ceases to be useful, when all these avenues of investigation are more or less exhausted and the theory does not provide much opportunity to discover novel facts, that we have what Lakatos calls a *degenerating* system. At that point, scientists start abandoning their allegiance to the old theory and begin to shift their allegiance to a competing theory that promises to provide new and more fruitful research avenues to explore. It is this process that eventually leads to a scientific revolution in which the old paradigm gets replaced by the new.

Other philosophers have similarly sought to find rationality in the way that scientists make theory choices. Larry Laudan says that scientific progress can be understood as one of ever-increasing problem-solving capacity.

Given that the aim of science is problem solving . . . *progress can occur if and only if the succession of scientific theories in any domain shows an increasing degree of problem solving effectiveness.* . . . We can say that *any time we modify a theory of replace it by another theory, that change is progressive if and only if the later version is a more effective problem solver (in the sense just defined) than its predecessor.* (Laudan 1977, 68, emphasis in original)

Laudan warns that ever-increasing problem-solving capacity does not imply that the process is leading toward the truth, saying "I do not even believe, let along seek to prove, that problem-solving ability has any direct connection with truth or probabilities" (Laudan 1977, 123). In other words, rationality in scientific practice and increased problem-solving capacity should not be conflated with questions of truth.

Similarly, Alan Chalmers tries to present science as an "objective and progressive enterprise" (Chalmers 1990, 8) even though he disagrees with what he calls the positivist strategy that aims "to defend science by appeal to a *universal, ahistorical* account of its methods and standards" (Chalmers 1990, 4, my emphasis). He instead says that the aim of science is more modest.

The aim of the natural sciences is to extend and improve our general knowledge of the workings of the natural world. The adequacy of our attempts in this regard can be gauged by pitching our knowledge claims against the world by way of the most demanding observational and experimental tests available. (Chalmers 1990, 115)

Philosopher of science Philip Kitcher goes further and argues that because science is self-correcting, it can actually distinguish truth from error. He goes into great detail to show that the eventual success of (say) Lavoisier's theory of oxygen over that of the earlier phlogiston theory can be justified as progressive on *objective* grounds, as bringing us closer to understanding what the "real" world is like. He offers what he calls a "broadly realistic view of science: scientists find out things about a world that is independent of human cognition; they advance true statements, use concepts that conform to natural divisions, and develop schemata that capture objective dependencies" (Kitcher 1993, 127). Philosopher of science and mathematics Hilary Putnam offers a similar view of realism and its value, saying that "a realist (with respect to a given theory or discourse) holds that (1) the sentences of that theory or discourse are true or false; and that (2) that what makes them true or false is something external—that is to say, it is not (in general) our sense data, actual or potential, or the structure of our minds, or our language, etc." (Putnam 1975, 69–70).

On the opposite end of the spectrum are the proponents of what is called the Strong Programme in the sociology of knowledge (such as David Bloor) who argue that science and mathematics knowledge are like every other field of knowledge in that they are *socially constructed* based on the experience of its practitioners, and that the idea that they occupy a privileged position in relation to the truth or the objective world is unjustified (Bloor 1991). He claims that what we call "knowledge" is what is *collectively endorsed* while what is held individually or idiosyncratically is mere opinion or belief (Bloor 1991, 5). While accepting the "reliability, repeatability, and dependability of science's empirical basis" (1991, 24), Bloor disagrees with the idea that those theories we consider true are so because they reflect an underlying objective reality and thus need no further sociological explanation, and that only those scientific theories that are now considered refuted need sociological explanations as to why they were once considered to be true. He says that those theories that we consider true are as conditioned by sociological factors as the so-called false ones, and should be subject to the same kind of examination (Bloor 1991, 7). Other sociologists of science such as Helen Longino do not go as far as Bloor and argue that just because scientific knowledge is socially and culturally constructed need not lead to "unbridled relativism" and that the objectivity of science can be reconciled with it (Longino 1990, ix, 221).

In their anthropological study of how scientists work, Latour and Woolgar arrive at the conclusion that "there are, as far as we know, no a priori reasons for supposing that scientists' practice is any more rational than that of outsiders" (Latour and Woolgar 1979, 29–30). In my opinion, such a statement goes too far. There are in fact good reasons for supposing that scientists *are* more rational at arriving at scientific conclusions, even if we cannot prove that their conclusions are true. While it may well be true that scientists are no more rational than anyone else when it comes to their *personal* lives and views, the *practice* of science *demands* that they use of evidence and logical reasoning to arrive at conclusions and that is as good a definition as any of what we mean by rational decision-making. Scientists are required *by their very profession* to have rational decision-making skills. It may be that people who already have those abilities are the ones who self-select to go into science. But even if they did not, they must necessarily acquire them during their long training and apprenticeship period, since otherwise that process will weed them out. Furthermore, the forging of a consensus view by the *community* of scientists, again using evidence and reasoning, reinforces rationality, just as in the legal system having many people involved in juries or panels of judges can justify the verdict being considered rational even if individual outcomes may turn out to be unjust and wrong.

The unavoidable presence of discrepant results and anomalies explains why the scientific community and those promoting anti-science agendas often speak at cross-purposes. Take the case of the theory of evolution. As with *any* theory at *any* given time, there are facts that are not explained satisfactorily by the theory. Scientists, tacitly following the negative and positive heuristics of Lakatos's sophisticated methodological falsificationist model, look for all the other possible factors other than the basic theory to see where the problem lies. This practice constitutes the "normal science" that Kuhn says that most scientists are engaged in at any given time and these efforts almost always result in eventually resolving the anomaly, though as we have seen, sometimes the process can take considerable time. But critics of evolution, often citing Popper's naïve falsification model, view these anomalies as immediately disproving the theory. They do not see the operation of these heuristics as the standard way that normal science operates but instead as a conspiracy among scientists to cover up the fatal weaknesses of the theory and delay its exposure.

Understanding this strategy of anti-science critics is essential if we are to effectively counter them. These critics zealously collect anomalies and are adept at pointing to one after another and demand that a defender of science explain away each and every one and, if we cannot, claim that they prove that the theory is false. Rather that trying to explain away each specific anomaly (difficult to do because no one person has encyclopedic knowledge of the context of all the anomalies in science even in their own fields of study) supporters of science should simply point out that the existence of anomalies by themselves *prove nothing at all about the validity of scientific theories*. Instead they should demand of the critics that what they need to do is *provide viable alternative theories that make predictions that can be tested and are supported by a preponderance of credible affirmative evidence in their favor*, because that is how the merits of competing theories can be compared and adjudicated.

The choice between competing paradigms is *always a matter of judgment* by credible experts in the field based on the preponderance of evidence, and is not forced upon us by this or that piece of data. The process by which an old paradigm gets replaced by the new, the time taken for the transition, and the degree of acceptance of the new paradigm can all vary considerably, with some old paradigms never quite going away, as the next chapter will discuss.

19
Why some scientific controversies
never die

[This chapter looks at why achieving consensus in science can be slow and getting unanimity of views on some scientific questions is almost impossible, because those who are determined to find ways to preserve their beliefs can always find ways to do so.]

Even when scientific experts are agreed that a theory about some phenomenon is the correct one, they will still concede that their conclusions are provisional in nature and subject to change. But the more encompassing a dominant paradigm is the harder it becomes to change it. The currently accepted value for even a seemingly simple question like the age of the Earth that was discussed in detail in chapter 6 has had many answers over time and the current answer of 4.54 billion years is the product of a *consensus judgment* that requires consistency across the fields of geology, paleontology, chemistry, physics, biology, and astronomy, and of theories in subfields within each discipline.

One of the consequences of the way science has evolved has been this kind of increased *intermeshing* of diverse fields, with theories in one area being woven with theories in others. It is this interconnectedness that gives much more strength to the conclusions than would be the case when just a single or a few theories are involved, analogous to the way that individual threads when woven together make a fabric that is much stronger than each thread alone. While much of scientific research involves working within a narrow area to explain new experimental results or observations or reconcile data with theory, others involve trying to bridge different research paradigms or resolving conflicts that occur in the areas of overlap of two or more theories. The way that diverse fields had to come together to arrive at a consensus on the age of the Earth is a prime example of that.

Note that not everyone will accept the consensus. Individual scientists or small groups of them might well continue to work on theories that have been rejected by the majority, hoping that they can marshal enough evidence to convince their colleagues that they are right. But they will be doing so on the fringes of the scientific community, largely ignored until such time that they can make a very strong case. It is important to realize that since one can never *prove* them wrong, such people might well be around forever, convinced of their rightness, with only the size of their numbers varying depending on the strength of the beliefs that drive their search.

A good example of the way that minority viewpoints in science can remain for a long time is the phenomenon known as *cold fusion* that has stuck around for nearly three decades, long after scientific consensus was reached that there was nothing significant going on there. Nuclear fusion reactions in which protons (the nuclei of hydrogen atoms) fuse together to form helium result in the release of huge amounts of energy, far more than chemical, mechanical, electrical, or other means of non-nuclear energy production. Such nuclear fusion reactions are the main energy sources that power the Sun and other stars. Since hydrogen is present in every water molecule, the ability to have controlled fusion reactions has the promise of providing an effectively unlimited energy supply. But in order for the fusion reaction to take place, two protons have to get close enough so that they overcome their mutual electrical repulsion due to the fact that they are both positively charged. Clearing that repulsive barrier to achieve this proximity requires the protons to be moving at very high speeds and thus have very high energy, that is, to be "hot." While the high temperatures that exist in stars is sufficient to achieve this, efforts to create fusion reactions on Earth involve constructing expensive apparatus such as powerful lasers or tokamaks (a device that uses strong magnetic fields to confine highly energetic charged particles in a small region) to simulate high temperatures. These attempts have made progress but have not as yet met the ultimate goal of being able to produce a sustained amount of energy output that is greater than the energy input. There are major hurdles still remaining because of the immense technical difficulties that need to be overcome and the costs involved.

The reason that the announcement of cold fusion created such a sensation in 1989 was because researchers Stanley Pons and Martin Fleischmann, both respected chemists at major universities, claimed to have found a way to enable the fusion reaction to occur at *room temperatures* using relatively cheap and simple apparatus. If true, their work would have revolutionized not only basic science but also everyday life by providing almost unlimited amounts of cheap energy. But many scientists were skeptical of their claims because the physics of fusion seemed to be well understood and precluded the kind of reaction that the two cold fusion proponents claimed to have seen.

The scientific community split into the three groups that typically form when a revolutionary finding is first announced. There were those who believed that the new claims were valid or at least plausible and that fusion may well be taking place. Because of the scientific importance of the effect and its major commercial implications, they felt that this was an exciting breakthrough that they wanted to be part of, and proceeded to try and replicate the events and generate corroborative evidence. This group consisted of those members of the scientific community who had expertise in related areas of science and were in a position to drop their other work and shift to this area. The second group consisted of those who were skeptics and they argued that something else, not fusion, was generating the heat and they set about making their case for alternative explanations. And finally the rest, the largest group by far, consisted of those whose own research was unaffected by whichever theory turned out to be correct and who watched with interest from the sidelines but did not take active part in the debate, instead continuing their own research in the belief that the other two groups would eventually sort things out between them.

In examining the cold fusion phenomenon, scientists looked at various aspects of it. There were experimentalists who tried to replicate the original results, because reproducibility is one of the most important features of science. The word "science" is derived from the Latin *scientia*, which means knowledge based on demonstrable and reproducible data. Experimental results that are not replicable under controlled conditions are simply not taken seriously which is why reports of paranormal phenomena get short shrift from the scientifically minded. Other scientists looked to see whether they could find corroborative evidence for the postulated fusion mechanism. For example, if standard fusion reactions were indeed driving cold fusion they should, in addition to the observed heat, result in the production of specific quantities of other particles that could and should be detectable. But such particles were not seen. As we have discussed, such negative results are rarely dispositive because one can always invent auxiliary hypotheses to explain them away and supporters of cold fusion tried to see if tweaking existing theories could generate new types of reactions that would allow for the fusion reaction to occur at low temperatures without producing the usual byproducts, thus explaining the negative results. Those efforts have not been fruitful but they continue.

As a result of the intense interest, many scientists studied the issue and fairly quickly, in fact within a year, the considered opinion of the majority of the scientific community was that whatever was producing the copious amounts of heat, it was not a fusion reaction at all but something more mundane that could not serve as a source of a revolutionary new form of cheap energy. Because it did not seem possible either theoretically or experimentally to substantiate the claim of cold fusion, many researchers abandoned this area and went back to whatever they were doing before.

But not all. Because of the tremendous benefits that would accrue if cold fusion did turn out to be true, some researchers are loath to completely abandon the field and there is a small group of dedicated scientists who keep the flame alive. They continue to do work on it and hold conferences using less sensationalistic labels such as Low Energy Nuclear Reactions (LENR) or Condensed Matter Nuclear Science (CMNS), perhaps because the term "cold fusion" has become seen as pejorative. Some of this work makes it into mainstream journals or conferences because what they are doing is real scientific research that is exploring aspects of fusion.

So is cold fusion real or spurious? While no definitive conclusion has been arrived at, what is true is that the consensus view of the mainstream scientific community is that there is no nuclear fusion involved. How they came to this conclusion will be discussed later when we look in more detail at how the scientific community reacts to challenges to its basic paradigms but the relevant point here is that believers in cold fusion have not disappeared.

This kind of persistence of minority views in the face of majority skepticism is not uncommon in science. What disappears over time, however, are *major debates* between the dominant and minority groups, something that occurs frequently in the early days of a new theory. Each group eventually ends up largely working in isolation from the other. Scientists are saved from endless sterile discussions about which theory is true because the Duhemian collective "good sense" of the scientific community can arrive at verdicts *based on the preponderance of evidence* as to which theories are worth working on, and these verdicts are widely accepted.

In adjudicating the truth or falsity of theories this way, the community of scientists act like a panel of judges in a court case (or a panel of doctors dealing with a particularly baffling set of symptoms), weighing the evidence for and against before pronouncing a verdict, once again showing the similarities of scientific conclusions to legal verdicts. But in the case of science and unlike in law, these verdicts are not, and can never be, *formal* statements by a designated adjudicating body because the way science works is too diffuse and anarchic for such structures to exist. These verdicts are arrived at *informally* but are powerful nonetheless. Word quickly spreads through the scientific grapevine that certain fields of research are not good areas to work in and young scientists tend to heed such advice from senior scientists and avoid them in order to avoid jeopardizing their career prospects. In fact, such informal decision-making systems can be even more powerful in forging a consensus than formal ones because there are no mechanisms by which to appeal the verdict, since the verdict itself is arrived at by *tacit* rather than explicit agreement. Supporters of a marginalized theory have no recourse other than the painstaking path of accumulating evidence sufficient to attract individual scientists over to their side until the consensus shifts, which becomes increasingly hard to do because of the absence of the energy that younger researchers with fresh ideas bring to any field.

In the case of cold fusion, its challenge to the *status quo* was quickly rebuffed and scientists reverted to the old paradigm because they did not see the preliminary evidence as sufficiently compelling to lead them to abandon it. But as we saw in the case of the age of the Earth, sometimes the impact of a new discovery (in that case radioactivity) takes the scientific world by storm and the community quickly changes its belief. In such cases the scientific community *as a whole* often changes its views faster than individual scientists. Within the span of just a decade, consensus estimates of the age of the Earth went from about 20 million years to over a billion years, a huge shift by a factor of 50. In that case, it was the status quo that was upended but again, not everyone agreed with the changes. This does not detract from the reputations of the individual scientists who held out against the change, however. Kelvin may have been mistaken about his estimates for the age of the Earth but he is still considered one of the giants of physics. Indeed, in addition to his groundbreaking work in many fields, his application of rigorous scientific methods and reasoning to problems that lay outside their immediate domain, and especially his emphasis on the need for consistency among the many theories that impinge on any question, led to major advances in the way science is practiced and his honored place in scientific history is secure.

The difference in the way that new theories of the age of the Earth and cold fusion were responded to does illustrate that the distinction between those new theories that are successful in displacing the old and those that fail to do so is not sharply defined but depends to a large extent on the persuasiveness of the arguments and evidence marshaled by either side. In the case of cold fusion, those who believe that there is something there are still largely working within the framework of science, trying to tweak the standard paradigms to see if they can get it to accommodate their effect without contradicting all the other theories involved. They have a major challenge on their hands because in order to get cold fusion to be viable requires significant advances, such as the invention of new fusion mechanisms. But it is not totally inconceivable that they might succeed because the changes required lie largely within a relatively narrow spectrum of science, mainly nuclear and particle physics, and may not require major changes elsewhere.

On the other hand, those who think the Earth is 6,000 years old have a much more difficult row to hoe in persuading others, even though theirs was the dominant view until just a few hundred years ago. It is not simply that the value they have for the age of the Earth is so vastly different from the scientific consensus. It is because we no longer live in an era where scientific disciplines work with some degree of independence from one another and where a result in one area could be challenged without causing upheavals in other areas. When calculating the age of the Earth depended on using just the Bible and other texts, the discovery of a new manuscript, the appearance of a new translation of an ancient document, or a new interpretation based on a different textual analysis could result in new values for the age. Other areas of knowledge did not provide strong constraints. But now that would not be enough. Unless all the other fields of study such as biology, geology, physics, chemistry, and paleontology adjust their theories accordingly to be consistent with this new development, it will remain an outlier and not become part of the scientific consensus. For this reason, we are unlikely to see a major shift in values for the age of the Earth.

Nowadays we realize that a scientific "fact," although identified with one particular field, is the product of a *network* of theories encompassing many fields that were once separate. To get an authoritative answer to the question of the age of the Earth one would probably go first to someone identified as a geologist or Earth scientist, but the answer she gives will not be the product of just geological research alone but of a whole complex of theories that spread far and wide across science. To change a fact in one discipline requires modifications in other areas as well. As we saw, arriving at the consensus view that the Earth is billions of years old required agreement with many areas of science. To now reject such a fact without exploring the consequences it has for all the other elements that went into its production, and addressing those contradictions satisfactorily, *is to reject science altogether*.

To be fair to the young Earth creationists, some of them do recognize this. If one delves into the field that has adopted the label of "scientific creationism" one will find quite imaginative ways that they have developed to respond to their critics so that they can justify to themselves their belief in a young Earth. In order to preserve their theory, they have introduced auxiliary hypotheses that modify the applications of theories of relativity, quantum mechanics, cosmology, geology, and paleontology. But those strategies have a decidedly ad hoc flavor, since their wider ramifications are not explored. What they do is take the objections to their young Earth individually and suggest an alternative mechanism to combat just that one, rather than take the comprehensive approach that is required. While the adoption of ad hoc mechanisms has been successful in the past in science, especially in the case of quantum mechanics, this was because they were followed up with more comprehensive theories. Young Earth creationists have failed to do that.

For example, consider the fact that we receive light from galaxies billions of light years away. How do young Earthers reconcile that with a universe that is a few thousand years old? One solution they propose is to say that their God, for some inscrutable reason, created the universe with those light beams already in motion and this misleads us as to how long they have been in transition. While such a claim is irrefutable, it is so clearly contrived to be self-serving that it is not palatable even to some creationists. Another proposed solution that sounds more scientific is to say that the speed of light has not been the same since the beginning of the universe. A third suggested solution is that time itself could have been flowing at different rates (Lisle 2007).

While these approaches address the specific observational problem of light from galaxies, they do so in a piecemeal way that does not explore the consequences of each assumption in other areas. A varying light speed or flow of time would have major implications for *all* of cosmology and many other areas of science, and these effects are not adequately taken into account, if at all. In short, young Earth creationists do not construct an alternative theoretical structure that is *consistent with other scientific facts and theories*, and this is their major weakness.

In science, every proposed solution to a problem also makes predictions for what should be seen in other areas, as we saw with cold fusion. This is how corroborative evidence is produced in support of a new theory. If one goes to a scientist with a novel solution to a problem, almost the first reaction will be along the lines of "If your solution to *this* problem is correct, then it should predict a particular result for *that* situation." This dialectical relationship between proposed solutions and predictions that can be tested is integral to science, and this making of connections and testing of predictions has to be carried out if one is to be taken seriously. You can see this with a *scientific* suggestion that the speed of light may have been much greater at the very earliest stages of the universe (Afshordi and Magueijo 2016). But unlike in the case of the superficially similar proposal to salvage the idea of a young Earth, this proposal, in addition to seeking consistency with other known facts, also predicts results that could be measured to test their idea. Also, since their greater speed of light only persists for an exceedingly short time at the very beginning of the universe and long before the formation of stars and galaxies, it is quite different from what the young Earth creationists propose and would not meet their needs, since the universe remains billions of years old.

The problem that young Earth creationists face is that the accommodation they seek requires the wholesale jettisoning of major accepted theories of science, all of which point toward a very old Earth. There is no way that the scientific community is going to undertake such an upheaval without being confronted with an overwhelming need. They simply do not see the evidence for a young Earth as credible enough to do so and hence young Earth creationists have been relegated to wither on the fringes of science or are considered to have rejected science altogether. But it is also the case that it will never be possible to convince those who believe in a young Earth that they are wrong. Whatever counterfactual instance one can show them, they will *always* be able to generate auxiliary ad hoc hypotheses that will explain it away and preserve their belief. If one believes in something strongly enough and one is determined enough, *it is always possible to rescue any theory or "fact" by means of constructing appropriate auxiliary hypotheses to salvage that belief*. To quote Jonathan Swift: "Reasoning will never make a Man correct an ill Opinion, which by Reasoning he never acquired" (Swift 1721, 27).

For an amusing example of how it is possible for the adherents of any theory, however preposterous it may seem to most people, to manufacture rejoinders to criticisms, go to the website of the Flat Earth Society (https://wiki.tfes.org/Frequently_Asked_Questions) and see how they defend that proposition. It may be that some members do not really believe that the Earth is flat and are just doing this for laughs, but that is beside the point. What is interesting is how they can construct an elaborate system to sustain an idea that almost everyone nowadays would dismiss as manifestly false.

Another reason for the durability of some ideas is because any determined individual or group could argue that different criteria, those that favor their own paradigm, should be used for adjudication between theories. This would be similar to an argument about whether the top batter in baseball is better than the top batter in cricket. Such an argument can go on forever with each side using different criteria in support of their position. This is why arguing with a determined believer is almost always futile, at least in the short run. The best one can do is to state one's case as best as one can and then walk away, hoping that the evidence and arguments that one presents will plant seeds of doubt in their minds that will take root and result in a change of view over the long run.

Many of the dissident groups from mainstream science genuinely believe in their alternative theories and pursue honest research in support of them, as in the case of cold fusion or, in earlier times, with theories of the aether or phlogiston. But the larger community of scientists, once they have reached a consensus judgment as to which paradigm they will work within, tend to ignore these alternative theories. While these theories quickly fade from the public eye, they only truly cease to exist when their most committed advocates give up the cause or die.

But sometimes there are strong, vested, *nonscientific* interests challenging the accepted scientific paradigm that can prolong the life of these discredited theories. Those interests exploit the feature that scientific theories are underdetermined by the data to deliberately undermine the scientific consensus in pursuit of *nonscientific* agendas and throw massive resources into doing so. We have seen this done by the tobacco industry when it came to the dangers of smoking, by the fossil fuel industry when it comes to greenhouse gases and climate change, and by religious creationists when it comes to evolution. Such people try to create the image that there is a state of permanent controversy in each area by throwing up one alternative explanation after another to the scientific consensus, despite each one getting shot down.

When it comes to climate change, for example, skeptics have proposed a variety of alternative theories, successively edging closer to the scientific consensus as each one is thoroughly debunked, but not willing to fully accept that consensus. Starting with the position that the Earth is not warming at all, they moved on to saying that it *may* be warming now but that this is part of a natural cyclic process of warming and cooling and that if we wait it out, the warming problem will disappear by itself. When that position became untenable, some then proceeded to say that irreversible warming *may* be occurring but is not due to human activity such as increased production of greenhouse gases but to factors outside our control such as solar activity, and so nothing can nor needs to be done. Some even go so far as to accept that the Earth is warming due to human activity but that this is good for life on Earth. Some religious groups such as creationists in the United States throw in a different argument. They believe that following Noah's flood, God gave us a promise that he would not destroy the world and so we are safe and nothing needs to be done.

As more and more evidence emerges that climate change is real, serious, and human-caused, each of these alternatives is brought in to challenge the latest scientific consensus and to try to impress upon the public the idea that the science is not settled and that nothing needs to be done. And as long as they have the money, resources, and other forms of support from the fossil fuel industry as well as from others who want to restrict or even eliminate government controls on polluting industries, they will continue to be a visible presence since money buys access to the media and enables them to keep these alternative theories in the public eye, however much the scientific community has rejected them.

This constitutes another, more unfortunate, similarity with the legal system because it is often the case that those who have plenty of financial resources can mount a much more vigorous case than those who do not. They can hire their own experts, examine the opponent's case in the minutest detail to find weaknesses however trivial, and produce mountains of alleged "counterevidence" that challenge each aspect of the evidence against them every step of the way. By using all these means, they can delay and obfuscate the process, giving the misleading impression that their case is much stronger than the evidence actually warrants. In the case of those who wish to discredit the scientific consensus on a topic, they will fund their own experts to give talks and write articles that oppose the scientific consensus, create think tanks to publish reports and newspaper op-eds, and provide forums for their in-house "experts" to get more exposure (Oreskes and Conway 2010; Supran and Oreskes 2017). Even though their work rarely makes it into the peer-reviewed literature because they do not meet the standards for scientific acceptability, these groups can create enough publicity through press releases and industry sponsored media events that the general public may think that there is a genuine scientific controversy when there is nothing of the sort. (Supplementary Materials G looks at how this latest effort against climate change warnings follows the same template as earlier efforts to discredit expert scientific consensus about the harm caused by tobacco use, acid rain, chlorofluorocarbons, and strategic missile systems.)

Such people also use the periodic reversals in science, which as we have discussed is consistent with normal practice, to undermine confidence in science.

Such reversals happen quite often in the field of medicine where the public has been confronted with one switch after another from the medical science community. Our understanding of the way that the human body works has changed dramatically from the days of the early physicians millennia ago. Even within our own lifetimes, we have encountered many reversals about what is a healthy practice and what is not, what are the causes of diseases and what are not, from hailing oat bran as a cancer preventer and red wine as reducing heart disease to downplaying their benefits later, from a stage of demonizing eggs and butter and cholesterol to later becoming more ambivalent about them, and so forth. Those with an anti-science agenda have exploited these periodical reversals to assert that scientists are either poor judges of truth or are even dishonest, and thus claims that their theories are true should not be taken seriously. These reversals can create a sense of cynicism among those who do not understand how science really works, leading them to believe that scientists' conclusions are not to be taken seriously.

Part of the problem we face is that the media trumpets each new result that emerges from scientific research laboratories as if it were certain knowledge, rather than provisional and subject to revision as new data emerges. Hence any subsequent reversal is also treated as big news, an upending of conventional wisdom, and brings with it new certain knowledge. These sensationalist presentations of reversals have been exploited by the enemies of science to argue that science is a form of art and its conclusions are mere opinions.

The reason for this misleading image is that most new research findings that make it into the major media come, for obvious reasons, from the fields of health and medicine since these are of immediate interest and concern to people. But a human body is an extraordinarily complex system. Furthermore, it is hard for both ethical and practical reasons to do controlled experiments on living systems. Medical researchers have developed methods, such as large-scale, randomized, double-blind clinical trials in which many confounding variables have been controlled for, that enable them to make judgments about causality from correlations in those cases where controlled experimentation is not feasible (Hill 1965; Jenkins 2004). But despite this, the knowledge that emerges from the field of medicine is necessarily less certain than those produced in fields such as physics that are not subject to the same stringent ethical constraints. Hence one should not be surprised by such periodic reversals.

The difficulty of doing research on living systems has raised questions as to whether medicine (and the social sciences) should be treated as more of an art than a science. Such questions miss the point. It is not the case that the conclusions of some fields are doubtful while we can be sure of others. The difference between medicine and fields such as physics is only one of degree. Even in those fields that deal with inanimate objects that can be subjected to much more rigorous tests, major reversals can and do occur. Ptolemy's geocentric model of the solar system was thought to be true for millennia before it was replaced by Copernicus's heliocentric model. Similarly, Newton's laws of motion and gravity were thought to hold true under all situations for centuries before they were found to fail when dealing with situations that lie outside the range of our everyday experience. The earlier theories were consistent with the observations that could be made with our naked senses or with more primitive equipment, but the development of technology that greatly extended the range of our senses resulted in new data that required major revisions or new theories.

Such reversals were major ones that impinged on vast areas of science and knowledge, and were so dramatic and well publicized that they have entered the public consciousness. But reversals are not really all that rare in *any* field of science. Deep inferential knowledge, things that lie far outside our immediate direct sensory world, forms the major component of our knowledge and we should expect it to change as technological advances enable us to generate more and better data, and the ability to make inferences improves. Such changes occur quite frequently within the many subfields that make up the overall framework of science but since those changes often do not directly influence many other scientific subfields, let alone reach the consciousness of the public at large, the ubiquity of this phenomenon is not recognized.

But such reversals do highlight one of the paradoxes of science. How can a theory pass tests of validity and thus be considered by scientists to be true and then later be discarded as false? And how can it be that this happens *repeatedly in every field and subfield of science*? Does that not imply that the way we determine the truth and falsity of scientific theories is itself flawed?

Consider a physician who is certified to practice medicine. Such a certification implies that the physician is competent to make diagnoses and the suggested treatments can be trusted. But on occasion, physicians often do have their credentials revoked because they obtained them by fraud or for malpractice or for gross incompetence. People can accept that this can happen occasionally, that no system of certification is perfect, and that there will be people who slip through the cracks. But repeated frequent occurrences would imply that either the certifying bodies were incompetent or that the certification standards were inadequate.

In the case of scientific theories, the standards are expected to be much higher, providing little or no room for error. We depend on scientific theories for much of our daily life and thus the goal should be to have in place a system to validate theories so that we can be very confident about them. The repeated reversals in the history of science suggest that at least in the past we were not successful in creating such a system. Does that mean that the scientists of the past were somehow inferior to those of the present generation and that our current theories will be far more robust? That seems unduly hubristic. As has been shown earlier, it was realized a long time ago by philosophers of science that we could not prove that any theory was "true" in any absolute sense of the word and that all knowledge is provisional and subject to change. This is not understood very well by the general public (and even by some scientists) and hence it is unsettling to them when the scientific community declares that a theory that once was considered true is now false, or at least not known to be true.

Some scientists have tried to counter this by taking the unfortunate tack of attacking the messenger, charging that epistemologists are undermining confidence in the truth claims of science. They have taken aim at all those who have had the temerity to challenge science's claim to revealing objective truths. A 1987 article in the prestigious science journal *Nature* severely attacked philosophers of science Thomas Kuhn, Karl Popper, Imre Lakatos, and Paul Feyerabend, saying that their work was undermining public support for scientific research (Theocharis and Psimopoulos 1987). A book *Higher Superstition: The Academic Left and Its Quarrels with Science* by biologist Paul Gross and mathematician Norman Levitt broadened the targets of the attacks to all those in academia, labeled by them as "the academic left," whom they felt were deliberately seeking to undermine science (Gross and Levitt 1994).

These attacks perhaps reached their apex with the convening of a conference in June 1995 under the auspices of the New York Academy of Sciences with the title *The Flight From Science and Reason*, where an assemblage of scientists and some of their allies in the humanities delivered broadsides against critics from "science studies," the area that examines and critiques the knowledge structure of science. The virulence of the attacks can be seen in a report on the conference (Hoke 1995) and in the talk given by Mario Bunge, a professor of philosophy and head of the Foundations and Philosophy of Science Unit at McGill University in Montreal.

> Walk a few steps away from the faculties of science, engineering, and medicine or law, towards the faculty of arts. Here you will meet another world, one where falsities and lies are tolerated, nay manufactured and taught, in industrial quantities. Here, some professors are hired, promoted, or given power for teaching that reason is worthless, empirical evidence unnecessary, objective truth non-existent, basic science a tool of either capitalist or male domination, and the like. Here, we find people who reject all the knowledge painstakingly acquired over the past half-millennium.
>
> . . .
>
> This fraud has got to be stopped, in the name of intellectual honesty and social responsibility.
>
> . . .

I submit that the academic charlatans have not earned the academic freedom they enjoy nowadays. They have not earned it because they produce or circulate cultural garbage, which is not just a nonacademic activity but an antiacademic one. Let them do that anywhere else they please, but not in schools; for these are supposed to be places of learning. We should expel the charlatans from the university before they deform it out of recognition and crowd out the serious searchers for truth. (Gross, Levitt, and Lewis 1996, 108, 110)

While the intensity of these so-called science wars has cooled in recent years, the antagonism toward the philosophy and sociology of science still exists in some quarters of the scientific community and occasionally erupts.

In my view, taking this hostile approach is a mistake because those who do so tend to make stronger truth claims for scientific knowledge than is warranted. Such claims leave people (and science) wide open to challenges by the enemies of science, since those critics can easily point to reversals in what we consider to be true to suggest that we don't know what we are talking about. To counter the propaganda and deliberate obfuscation of those with anti-science agendas, the public needs a much more sophisticated view of how science works than is currently presented to them in formal education or informally in the media or by overzealous supporters of science like those mentioned here. It would be far better for the scientific community to take seriously the findings of the historians and philosophers and sociologists of science and use them to build a more accurate image amongst themselves and the public of the way that science actually functions. That resulting image would be more robust and a better defense against those who are anti-science, because it would enable the public to understand that despite periodic changes in what we think is true, science is far and away the most reliable creator of knowledge and that by rejecting it, we are rejecting the very basis of *all* knowledge.

While we have undoubtedly progressed in terms of the technology available to study nature and the more sophisticated methods we have developed (major innovations like randomized, large-scale, double-blind tests, advanced mathematics, and refined statistical tools have enabled us to eliminate many sources of error and bias that existed before), there is no reason to think that we have eliminated *all* the ways in which we could go awry or that we are more careful or more ingenious than our predecessors. It would be an extraordinary stroke of luck if we just happened to be living in a time when we have stumbled upon at least some true theories even though we cannot prove them to be true.

Even though we have not *as yet* been able to do so, can we hope that some day we will find a way of unequivocally determining the truth and falsity of theories? Or is such a goal *intrinsically unattainable* and forever out of reach? The latter conclusion is a disturbing one that seems to fly in the face of experience, since we seem to have arrived at least at some scientific conclusions that seem to be incontrovertibly true. I once heard a scientist say in a talk that there are some inferential pieces of knowledge that must surely be independent of any theoretical construct. As an example, he said that water consists of two hydrogen molecules and one oxygen molecule. He said that he could not possibly conceive of how water could ever be shown to be anything else. Such examples can be multiplied many times. Surely the existence of the electron is a theory-independent fact? Surely too that the speed of light is independent of the state of motion of the source or the observer, a bedrock principle of physics that has led many successful predictions and practical applications? It does not seem, at least on the surface, that those things could possibly be false without calling into question the entire scientific framework that we now have.

The existence of electrons, the composition of water, and the invariance of the speed of light seem to be rock-solid facts. We are thus tempted to believe that the theories that led to them must also surely be true. But are they? How would we know? As I have argued, we cannot *prove* by evidence and logic that our theories are true. Hence it is safer to assume that our present day knowledge is also provisional and subject to invalidation in the future. Even if our present theories last for a thousand years, we cannot be sure that some day in the future, they will be replaced by new ones, and our present-day seemingly rock-solid "facts" may disappear along with them.

Is there any hope for establishing the eternal validity of *anything* that we believe to be a scientific fact? If all scientific facts are dependent on the validity of the theoretical structures that generate them, and we cannot prove those theories to be true, then is there nothing that we know for sure, at least when it comes to inferential knowledge? This leads us back to the primacy of theories in our knowledge structure and the importance that is assigned to finding theories that we think are valid and eliminating those that we think are of no use. But how do we make such judgments if neither truth nor correspondence with reality can serve as a measure? This will be discussed in the next chapter.

20

How science evolves and the Great Paradox of science

[This chapter finally confronts the Great Paradox: If science is progressing, what could it possibly be progressing *toward* if not the truth? It argues that the way that scientific paradigms evolve is analogous to the process of biological evolution, in that both are conditional on the environment that exists at any time and thus there is no reason to believe that the evolution of scientific theories is heading toward a unique truth. The strong impression of directionality is because scientific history in textbooks is reconstructed *after the fact*. Scientific evolution, like biological evolution, is not teleological.]

Philosophers and historians of science generally agree that science progresses and that this progress is manifested by the periodic shifts in allegiance of the scientific community as a whole from an old paradigm to a new one that is seen as *better* than the one it replaced. This shift is not done in unison or by fiat but by a much more informal process in which individual scientists, especially ones new to the area, start working within a new paradigm, winning over converts until it becomes the consensus of the community, while the adherents to the old paradigm get increasingly marginalized and eventually fade into obscurity.

Where there is disagreement lies in what causes these shifts. It is generally agreed that there is no clean distinction between theory and experiment. As a consequence, what we sometimes call "nature," through the results of observations and experiments, cannot play the role of an impartial and objective arbiter of theories, confirming those that are true and rejecting those that are false. Even assigning relative probabilities to theories turns out to be unworkable in practice. There seems to be an unavoidable element of subjectivity in how old paradigms get replaced by new ones. At any given time, which of the competing theories is most worthy of acceptance is a *consensus judgment* made by the scientific community using criteria that are not as clearly objective as many scientists would like to believe. What is at issue is whether that transition occurs because of the use of some kind of vague "good sense" as suggested by Pierre Duhem, by falsification as suggested by Karl Popper, by a change in perception akin to a gestalt switch as proposed by Thomas Kuhn, or by a more methodical process by which scientists move from away what is sensed to be a degenerating research program to a more progressive one as people like Imre Lakatos, Larry Laudan, and Alan

Chalmers argue. But none of those methods *guarantee* that the new paradigm is better than the old in any absolute sense and hence we cannot be assured that scientific theories are progressing inexorably toward the truth.

And yet the fact of scientific progress seems indisputable. Can anyone seriously doubt that our theories of today are much better than those we had five hundred years ago? If they are not, how can we explain the great strides that have been made in controlling and even changing our world? And if indeed science is progressing, what could it possibly be progressing toward if not truth, as philosopher of science Philip Kitcher and others argue (Kitcher 1993, 90, 128–33)?

This is the Great Paradox of science.

Kuhn made two highly suggestive analogies about the nature of science that illuminate the relationship between progress and truth. The first is one that I have discussed earlier in chapter 10(c) and that is that the process by which an old paradigm gets replaced by the new has parallels to *political revolutions* in that there is neither a set of objective criteria nor a neutral arbiter within an overarching framework within which the competing paradigms reside that can make the determination as to which one should determine the future direction of research. The final decision as to which paradigm emerges victorious is largely determined by the weight of evidence, but also by other factors such as the contemporary social and cultural environment and the persuasive and institutional power of the sides that are competing. A paradigm that manages to claim the allegiance of influential individuals and institutions and thus can influence the flow of resources and attract the next generation of scientists has a better chance of emerging as the victor.

As in the case of political revolutions, to the victors go the spoils. Just as political history is written by the winners to show that they were the virtuous ones with right on their side, so it is with scientific history. Those whose favored paradigm emerges victorious are the ones who rewrite the textbooks that teach the next generation of scientists and the memoirs that recount that history, and these reconstructions are invariably formulated to suggest that the victorious paradigm is the best and succeeded because it was right and the old one was wrong (Kuhn 1970, 138). The way it does this is by focusing attention on one or more experiments that supposedly falsified the old paradigm and suggesting that the questions that the current paradigm deals with the best are also the most important ones that have long been the subject of scientific inquiry, and that the reason that the older theories were rejected was because they failed to adequately answer them. Sometimes the discarded theories are spoken of in such a disparaging manner that the current generation of students may be mystified as to how they could ever have been taken seriously at all. Such whiggish historical reconstructions lead to the idea of science progressing becoming firmly embedded.

A good example of how older theories are deliberately discredited as part of the strategy to gain acceptance for new ones is seen in the case of alchemy. This was the search for the "philosopher's stone" that would enable the transformation of one type of material to another, especially base metals to precious ones, and also possibly lead to an "elixir of life." Most people now dismiss alchemy as absurd even if they know little more about it than its name. The fact that Isaac Newton spent so much time on this research is seen as an inexplicable aberration by this great scientist, perhaps indulged in during his dotage. But in reality, Newton did this work in the prime of his life because it was seen as important (Iliffe 2017). The experimentation by alchemists lasted from the Middle Ages until the end of the seventeenth century and played a central role in the progress of science by creating a large body of empirical knowledge that enabled the emergence of present days sciences such as chemistry, metallurgy, and pharmacology. Lawrence M. Principe describes how "alchemy's outcast status was created in the eighteenth century and perpetuated thereafter in part for strategic and polemical reasons," as part of the effort to gain acceptance for the new discipline of chemistry.

> It was spokesmen for scientific societies and institutions—like Bernard de Fontenelle and Etienne-Francois Geoffroy at the Academie Royale des Sciences and Boerhaave at Leiden—where chemistry was struggling to take on a new identity in terms of professionalization and social legitimacy, who led the charge. They cast alchemy as an intellectual taboo, its practitioners as socially unacceptable and disruptive, and its content and practice as something other than the chemistry they represented. This campaign was so successful that [alchemy] disappeared from respectable circles within a generation (although some of the most prominent eighteenth-century chemists who rejected it publicly continued to pursue it privately). (Principe 2011)

Scholars are now trying to rescue alchemy's reputation so that it is no longer the target of ridicule.

While Kuhn's analogizing of scientific revolutions with political ones is illustrative, there is one aspect of it that may be misleading. Political revolutions are often dramatic, usually take place over a fairly short time period, can sometimes be violent, and it is obvious to all that one has occurred. Although scientific revolutions as portrayed in textbooks may share some of those dramatic qualities (except for the violence), in reality many scientific revolutions may take a long time to be completed since they involve the slow shifting of allegiances, and as such they may be invisible during the time when they are occurring even to those involved in them, only recognizable as revolutions in hindsight and that too much later.

The second of Kuhn's insights is that the evolution of scientific ideas is analogous to the evolution of life. According to a simplified version of Darwin's theory of natural selection, changes in organisms occur through random genetic mutations and there are potentially a large number of mutations that can happen. Not all of them will occur and of those that do, only some will grow in numbers and end up dominating the population while the others die out. Success occurs because of genetic drift or because the mutations represent a favorable adaptation (or are linked to such an adaptation) and thus have some preferential advantage for survival in the environment that exists *at that time*. This process gets repeated over time as various species come into being and disappear.

But while the *process* of evolution is well defined, it is not teleological, that is, *goal-directed*, in any way. In other words, evolution is not progressing *toward* a particular goal that was preordained from the time when the first self-replicating molecule came into being and the process began. There is no scientific reason to believe that human beings and all the other living things on our planet right now were the ultimate goal of the evolutionary process that began in the so-called primordial soup. Although there were an infinite number of possible evolutionary chains that could have evolved from the initial conditions, it just so happened that the chain that actually occurred was the one that led to the present state. What we have now is just the accidental byproduct of a process that could have gone in many different directions. It is quite conceivable, indeed likely, that if we ran the clock again we would arrive at a different end point that did not include humans, although we may not be able to envisage the final outcome in all its details. The reason that this particular chain occurred out of all the possible ones was due to both chance as well as the complex interplay between each organism and the environment *at any given instant of time*, since that is what largely determines how any particular organism would evolve at that time.

Even making a clean distinction between an organism and its environment is an oversimplification, similar to the error in thinking of theory and experiment as completely distinct. All organisms are also part of the environment. Just as theories are unavoidably embedded in experimental results, the organism and the environment are not distinct entities but are inextricably linked because the environment is not an entirely static and passive bystander whose role is to determine which mutation will survive and which will die out. To understand this, it is helpful to make a distinction between what I will call the *hard* environment and the *soft* environment. The hard environment consists of those factors that have been largely unchanging over the course of biological evolution, such as the effects of the Sun, the Moon, and the Earth's gravitational field. The soft environment consists of the more immediate environmental factors that have changed over time, such as the composition of the atmosphere, climate, and biological features such as the flora and fauna that exist at any time.

Organisms are part of the soft environment and when any organism changes, the soft environment changes too so that with each new stage of evolution, organisms encounter a different soft environment from their predecessors. So while both hard and soft environments determine how an organism evolves, the organism in turn determines how the soft environment will evolve. Rather than a purely hierarchical relationship in which the soft environment rules over the organism, it is a dialectical relationship, each influencing the other. In the modern age, it can be argued that the organisms we call human beings are changing the environment quite dramatically with their extensive use of fossil fuels and the depletion of rain forests and polar ice caps. Certain species of wildlife are now either extinct or near extinction due to the impact on their habitats of actions taken by humans. The effect is so dramatic that some have even coined the term *anthropocene* to label the current human-influenced epoch. Although it is hard to predict what this pattern of human behavior will do to future evolutionary changes, it is safe to assume that it will have a different effect than if we decided to reverse course in all these areas of environmental impact.

Thus the question of which of the many possible mutations of an organism will survive and flourish is not a question that can be answered once for all time. The answer will vary depending on *when* the mutation occurs. A mutation that had no chance of survival in one era could well be the most favored one if it had occurred a million years earlier or later when it might have encountered a very different environment. It is this time- and context-dependent complex dialectical relationship between organism and soft environment that makes it hard to believe that evolution is goal-directed, that the world around us now is the *inevitable* consequence of a process that started billions of years ago and proceeded according to natural laws.

One way to retain a goal-directed view is to invoke a supernatural being who took away the element of chance by intervening at appropriate times in order to supersede the natural laws and thus ensure that the state of life that we have now is what it was preordained to be. Failing that, it may be possible to argue that humans were destined to emerge because the unchanging *hard* environment swamps the effects of the soft environment to the extent that it determines the entire process of evolution. It is true that the features and range of biological development are *constrained* by the laws of physics that are believed to be unchanged from the beginning of the universe (Cockell 2017). This may explain the phenomenon of *convergent* evolution in which similar features evolved independently multiple times in different forms. Eyes may have been destined to appear because the constant presence of sunlight provides immense benefits to organisms that can see and thus creates great selection pressure for eyes to emerge from the evolutionary process. Thus if the evolutionary clock were run again, it would not be surprising to see creatures with eyes in some form, though the details of their construction may differ in many ways from what we have now. Similarly, the constant presence of an atmosphere would provide a hard environment that enables flight so that that ability that we now see with insects and birds may also have been destined to emerge. The streamlined bodies of fish may have also been destined to appear to enable smooth passage through water.

Using Kuhn's idea of using natural selection as the analog for the evolution of scientific knowledge, the success of any new scientific theory in gaining acceptance depends on the *scientific environment* that exists at the time that the theory is created. If the environment is favorable, the theory can flourish. If not, it may simply wither away. As an example, the heliocentric solar system is known to have been suggested by Aristarchus in ancient Greece as far back as 250 B.C.E. but his model was not accepted. But the same idea, introduced by Nicolaus Copernicus (1473–1543 C.E.) and published at the time of his death in his book *De revolutionibus orbium coelestium (On the Revolutions of the Celestial Spheres),* did change our view of the world and the influence of nonscientific factors in its emergence are worth examining.

Scientific and popular folklore says that the Copernican model, although clearly better than the Ptolemaic one, was resisted because of religious dogma. It is widely assumed that Copernicus's model was opposed because the Earth being removed from the center of the universe was seen as a demotion for God's chosen people. But history does not support that view. In his article *The Great Copernican Cliché,* Dennis R. Danielson points out that in Copernicus's time, the center of the universe was *not* considered a desirable place to be. "In most medieval interpretations of Aristotle's and Ptolemy's cosmology, earth's position at the center of the universe was taken as evidence not of its importance but (to use a term still in circulation) its grossness." In fact, it was believed by ancient and medieval Arabic, Jewish, and Christian scholars that the center was the *worst* part of the universe, a kind of squalid basement where all the muck collected. Medieval writers described the location of the Earth as "the excrementary and filthy parts of the lower world" and that we humans are "lodged here in the dirt and filth of the world, nailed and riveted to the worst and deadest part of the universe, in the lowest story of the house, and most remote from the heavenly arch." Cardinal Bellarmine in 1615 said that "the earth is very far from heaven and sits motionless at the center of the world" (Danielson 2001).

Danielson points out that in Dante Alighieri's (1265–1321 C.E.) *Divine Comedy,* hell itself is placed in the inner core of the Earth, which is even closer to the very center of the universe, consistent with it being considered a foul place. Dante also speaks of hell in ways consistent with Aristotelian dynamics, not as full of flames (because fire would be up in the sky, displaced by the heavier earth) but as frozen and immobile. By contrast, heaven was "up" and the further up you went in the sky, away from the center, the better it was. So Copernicus, by putting the Sun at the center and the Earth in orbit around it, was really giving a *promotion* for Earth and its inhabitants, not a demotion, by taking them closer to the heavens.

As Kuhn says in his book *The Copernican Revolution*, the Earth was not believed even by Aristotle to *be* the center of the universe; it was thought to be *at* the center of the universe, as evidenced by him saying "It so happens that the earth and the Universe have the same center, for the heavenly bodies do move toward the center of the earth, yet only incidentally, because it has as its center at the center of the universe" (Kuhn 1957, 84). That is a crucial distinction. The geocentric Ptolemaic model was based on Aristotelian dynamics that placed the Earth at the center of the universe simply because it was (as was believed then) the most massive object around. There was nothing religious or even metaphysical in this reasoning; it was quite physical and naturalistic.

In Aristotle's cosmology, the universe was finite and the heavens existed beyond the outermost sphere. It was believed that there was a center of the universe (defined as the center of the large rotating outer sphere in which the stars were embedded) and that matter was drawn to that center. In this cosmology, the directions "up" and "down" were well defined. "Down" was toward the center of the universe and "up" was away from it, toward the sphere containing the stars. The elements were earth, air, water, and fire and each element had its natural affinity for a location in this universe. As could be seen from the fact that rocks fell to the ground, earth (being heavy) was drawn to the center. Flames leaping upward showed that fire (being light) was drawn toward the heavens. Water and air occupied their appropriate positions between these two extremes.

Furthermore, the Copernican model was not obviously superior to the Ptolemaic one. The geocentric model explained many things such as why objects fell to the ground when released from any point on the Earth's surface (because they were being attracted to the center of the universe) and why the Earth was spherical in shape. It also explained why the Earth was motionless at the center because for it to move, there had to be something that took it away from the center and no such agency was in evidence.

The major reasons for opposition to the Copernican model were quite secular.

His full system was little if any less cumbersome than Ptolemy's had been. Both employed over thirty circles; there was little to choose between them in economy. Nor could the two systems be distinguished by their accuracy. When Copernicus had finished adding circles, his cumbersome sun-centered system gave results as accurate as Ptolemy's, but did not give more accurate results. Copernicus had failed to solve the problem of the planets. (Kuhn 1957, 169)

Strictly speaking, there was no objective reason for preferring the Copernican model over the Ptolemaic one and Copernicus could have met the same fate as Aristarchus. But he did not and the reasons why are important.

> Judged on purely practical grounds, Copernicus' new planetary system was a failure; it was neither more accurate nor significantly simpler than its Ptolemaic predecessors. But historically the new system was a great success; the *De Revolutionibus* did convince a few of Copernicus' successors that sun-centered astronomy held the key to the problem of the planets, and these men finally provided the simple and accurate solution that Copernicus had sought. (Kuhn 1957, 171)

We see once again how the ability to persuade a few able people to work with a new idea is often the key to its eventual successful adoption by the wider community, provided that work proves to be productive. There was something in Copernicus's ideas that appealed to a few astronomers and these converts to the idea saw in it (in Lakatos's terms) a progressive system that promised fruitful new avenues of research, while the Ptolemaic system was seen as a degenerating system, losing steam and looking exhausted.

As discussed earlier in chapter 8, a key convert was the astronomer Johannes Kepler (1571–1630 c.e.), who realized that using elliptical rather than circular orbits in the Copernican system resulted in much better agreement with observations. His insight was made possible because of the more accurate astronomical measurements made by Tycho Brahe (1546–1601 c.e.), one of the greatest naked-eye astronomers of all time, that provided new and more accurate data while eliminating some earlier erroneous data that were leading people astray. What Kepler's innovative idea of elliptical orbits did, when coupled with his law of areas, was to enable the Copernican model to dispense with cumbersome and complicated epicycles. Kepler's *Rudolphine Tables* for planetary motion based on the Copernican model (published in 1627) were the most accurate for describing planetary motion. It was better agreement with data that persuaded other astronomers that the idea of a heliocentric system coupled with elliptical orbits was worth taking seriously and aided in its adoption. Incidentally, though Brahe's work was instrumental in leading to the eventual adoption of the Copernican model, he himself was an opponent of it.

But many astronomers felt that the Ptolemaic system, although complicated, could ultimately be made to work. So while they hailed Copernicus's model and used the resulting tables and methods, most were skeptical of his central idea of a moving Earth and of Kepler's elliptical orbits. They dismissed it as some kind of ad-hoc trick (similar to the way that Planck's quantum hypothesis was initially viewed centuries later) that turned out to be a useful tool for calculations. This idea that the motion described by a model is a convenient fiction was not unprecedented since Ptolemy himself had said that not all of his own epicycles had to be considered physically real and that some were to be thought of as merely mathematical devices that gave numerically sound results (Kuhn 1957, 186).

There were good reasons for astronomers at the time to view Kepler's idea of elliptical orbits warily as somewhat bizarre. Such orbits created all manner of new problems that were not present with circular orbits. When seen through the eyes of the people at that time, assuming circular motion was quite reasonable. Since there were no theories of force or gravity, astronomers needed to have an explanation of motion. In the case of circular motion the question could be answered using a hand-waving argument, by saying that it could be viewed as a natural state and an initial condition, that once an object had been set in such motion at the beginning of time it would continue to be in that same state and thus did not require any further explanation (Kuhn 1957, 245). If you had more complicated motions like ellipses, that would have meant that the speeds and distances of the planets from the Sun were constantly changing and this required a *dynamical* theory of motion that simply did not exist in those pre-Newtonian times.

The Copernican model also raised new problems. It required the Earth to be in motion but it did not say what caused it to move away from the center of the universe. Furthermore, since the Earth was still believed to be the most massive object in the universe, then if it was not drawn to a fixed point at the center of the universe, did that mean that there was no center at all? If there was no center to the universe, could that mean that the universe was infinite in size? If the Earth was not stationary at the center, but was midway in the sequence of planetary orbits around the Sun, then how could one define "up" and "down"? Why would objects fall "down" if the Earth were not at the center of the universe? How could objects that were thrown vertically upward fall back to the same point if the Earth were not at rest? It took the later work of Galileo and still later of Newton published in his *Principia Mathematica* in 1687, that sealed the scientific case in favor of Copernicus by putting it on a firm theoretical footing, by showing that elliptical orbits emerged naturally from the laws of motion and gravitation. Thus the Copernican revolution, far from being quick, required a period of 150 years for its full adoption.

It should now be clear why Copernicus's heliocentric model became the dominant paradigm while Aristarchus's suggestion of a similar system failed to do so. The reasons for widely differing responses to essentially the same idea is due to the scientific, technological, political, social, and religious climate that surrounds any idea at any given time, playing roles analogous to the soft environment of evolutionary theory. If that climate is sufficiently favorable, an idea will flourish and gain adherents. If not, the idea can fade into obscurity. The successful theory also determines the *future* scientific environment that later new theories will encounter. At each branch on its past road to the present (where a branch represents a period where the existent paradigm was perceived to be inadequate), the scientific community determined which of the competing scientific theories best solved the *immediate* problems confronting it, using the yardsticks available at that time. It then adopted that theory as the best and all other competitors slowly disappeared. But this very action also *changed the environment of scientific knowledge* in that it created a new paradigm centered around the victorious theory within which all new "mutations" (i.e., new paradigms) had to occur. The environment also changed the character of the scientific community, because it was the practitioners of the victorious paradigm who were now in control and thus set the framework for new research.

To make this important point more concrete, note that at any given time, there exists a set of observations that we call "experimental data." Although we have argued so far that experimental results cannot serve as an impartial and objective surrogate for nature because of all the auxiliary hypotheses involved in their interpretation, we will still use that phrase as a convenient label to describe the combination of raw sense-data, interpretive theories, and the unproblematic background knowledge that goes into it. The best theoretical framework at any particular time, which we have called the paradigm, is required to explain satisfactorily most of the existing data, or at least those that are considered the most significant. But it never explains *all* of them. Any given theory has an infinite number of predictions that can, in principle, be made within its domain. Similarly, there are an infinite number of experiments that can be performed. Scientists do not randomly select experiments or theoretical calculations to perform. They are not like children casually picking up stones in the hope that one will turn out to be a gem. Scientists have limited time and resources and need to pick and choose their activities very carefully. Only those theoretical calculations and experiments that are technically feasible, promise elucidation of the paradigm, or provide tests of it are attempted. Theoretical calculations that require tortuous analyses or the use of auxiliary theories that are themselves not well established, or experiments that no one knows how to properly set up or lack the funds or technology to perform, are typically avoided.

Since no theory ever explains *all* the data that falls within its domain, there will *always* be predictions of the theory that do not agree with experiment. This set of problems will form the focus of much of normal science. While many of these disagreements will be eventually resolved, from this set of unexplained data will eventually emerge the so-called anomalies that are resistant to solution within the rules of the existing paradigm and which eventually lead to the crises that result in the overthrow of the dominant paradigm, to be replaced by its successor. *It is here that subjectivity comes into play and prevents us from concluding that the scientific theories that we now have were destined to appear.* Figure 20.1

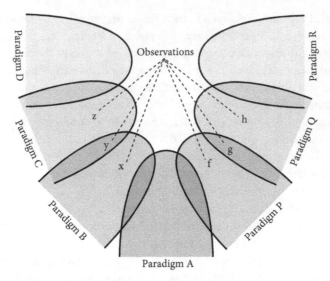

Figure 20.1 The Evolution of Scientific Paradigms
Credit: Josh White

illustrates in a simplified schematic way the contingent nature of the evolution of scientific paradigms.

In this figure, time flows upward. The hairpin shapes represent possible theories and their associated paradigms. The area encompassed between the arms of each hairpin represents the successes of that paradigm, that is, where its predictions are believed to be in agreement with the data. The hairpins are open-ended because there are potentially an infinite number of predictions any nontrivial theory can make. Only a small fraction of such experiments and calculations will actually be done or even can be done because of limitations of time, money, technology, or interest. Thus a paradigm may be deemed to be highly successful even though it can never be fully tested. The areas *outside* each set of arms represent those experiments that were either not done or were done and found to be in disagreement with the predictions of that paradigm, that is, anomalies.

Assume that the dominant paradigm at one time is represented by the shape *A* and that data points *x* and *f* do not agree with its predictions and hence lie outside the arms of *A*. These are considered anomalies. If the existence of these anomalies is confirmed by others and they are considered to be important consequences of the paradigm, strenuous efforts will be made to resolve the discrepancy, either by modifying the paradigm so that the arms of *A* spread wider to encompass the points or the experiments are repeated to see if the data points shift to within the arms of *A*. If one of them (say *x*) is seen as important and is highly resistant to the best efforts of the most capable scientists, a sense of urgency could arise and new theories will start to be invented and investigated to see if they can resolve the problem.

Suppose that another theory B is proposed so that the competing theories A and B have partial overlap in the sets of data that they explain, but neither completely encompasses the other, which is the normal situation in science. Let us assume that B explains the troublesome data point x. That will weigh in favor of B and against A. As Lakatos argues, the chances of acceptance for theory B will receive a strong boost if it makes a novel and risky prediction (in the Popperian sense) of a *new* result y that, if subsequently measured, turns out to be in agreement with the prediction of B but not with that of A. Such a success will heavily increase the chances of B becoming the new paradigm, especially if the prediction was a spectacular one that would not have even been considered under A. If more data points emerge that show this same pattern, then theory B can over time emerge as the new chosen paradigm, replacing A.

Note that it is not obvious that the new paradigm B is *better* than paradigm A in any objective sense. It does not explain all the data that theory A did and in fact it usually (at first, at least) explains *fewer* experimental results because it is new and relatively few people have worked with it. It may not even address highly successful earlier predictions of theory A, the ones that had led to A becoming the accepted paradigm in the past. Its claims to becoming the new dominant paradigm largely rest on it being partially consistent with some old data that were also explained by A, being able to explain the point x that had persistently stymied the efforts of those working with theory A and had been the cause of growing concern, and because it made some new predictions that were confirmed. There is no *objective* reason for preferring paradigm B over A and yet the history of scientific progress shows that this is how scientific paradigms evolve. When scientific history is written long after B's ascendancy, result x is treated as a *falsifying* experiment for A while y is treated as a spectacular new confirming prediction for B. Eventually theory B will evolve by the same sort of process and reach its own time of crisis created by a new anomaly z, be replaced by theory C, and then on to theory D, and so on. But this particular chain of theories depends on the nature of theoretical interest and the computational and experimental technology at the time of each paradigm choice *that began with the crisis represented by the point x* (Feyerabend 1993, 156–57).

As actual historical examples, in the transition from Newton's theory of gravitation to Einstein's, A would refer to the Newtonian paradigm, B to the Einsteinian paradigm, x to the anomalous behavior of the perihelion in the orbit of Mercury, and y to the prediction of light being bent by the gravitational field of the Sun. Similarly, in the transition from the corpuscular theory of light (represented by

A) to the wave theory (B), x could represent the speed of light being slower in water than in air and the bright spot predicted by Poisson would be y.

What of the other experimental anomaly f for paradigm A? This will fade into obscurity after paradigm B's ascendancy because it did not meet that paradigm's criteria for research worthiness, while x is elevated to a position of great importance as a crucial falsifying experiment. We have seen in chapters 10(b) and 18 that falsification is not viable in that no single anomalous event can falsify a theory and that theory choice is based on many factors and is not entirely objective. But textbooks are written from the perspective of the victorious paradigm B and treating x as an experimental result that falsifies A paints a narrative structure that makes the process seem objective and validates B. Thus falsification is essentially a myth-story, a rewriting of scientific history, but is done so routinely in textbooks that it has become perceived as playing an essential role in the progress of science.

It is quite conceivable that there could be another theory P that explains anomaly f but did not command the same degree of attention because, for whatever reason, the scientific community got focused on x, and P did not address that particular anomaly. Since neither A nor B encompass the point f, this conflict between theories A and B and experimental result f will not be perceived as serious, if it is noted at all. So the experimental point represented by f and the theory P never figure significantly in any discussion over paradigm choice during the time of crisis for A. Once paradigm B is accepted as the dominant one, the only problems it addresses are those that lie within its own range of worthiness and if this range does not overlap those of P, then theory P will never be considered again, *although it might be a very worthy theory.* Theory P is ignored simply because its areas of success were not considered important or interesting *at the time of crisis for theory A* caused by anomaly x.

If the context had been such that f became the cause of a crisis and the focus of attention, then the evolutionary branch that scientific knowledge would have followed would have been that from A to P and onward along that branch through Q, R, and so on. This would have been a completely different chain and eventually led to a different understanding of nature than the one we have now, or a different "truth," and f would have been classified as the falsifying event for theory A.

It is conceivable that the chain that began with B might eventually, at some later time, overlap with the chain that might have begun with P, as happened with the heliocentric theory that surfaced in 250 c.e., disappeared, and then reappeared in the sixteenth century when it succeeded. But there is no guarantee that this will happen and in practice it seems to hardly ever do so. What seems to happen is that the chains diverge and never overlap again and that even though P and its successors may, at a later time, be more suitable and fruitful theories than whatever is currently being used, they may never come to light. It is thus quite possible that there are many excellent theories that have never been born, or having been invented are yet doomed to obscurity purely for the reason that they could not address the *particular* questions that were of interest *at the time they were originated.* The fact that at any given time we have just a few theories competing for supremacy around a few anomalies may mislead us into thinking that we are nearing the truth, when all it may signify is the limited range of our attention, the paucity of our imagination, and the *strict constraints imposed on us by historical accident.* Just as biological success is based on what best meets the *immediate* environmental constraints that are historically contingent and not on what may be best in the long term, so it is with scientific theories.

We thus end up facing the ultimate paradox. Science works extremely well, as I've emphasized repeatedly in earlier chapters, and seems to be getting better all the time in that, thanks to it and the many technologies associated with it, it has revolutionized our lives and enabled us to do things that would have been un-imaginable even just a century or two ago. This level of immense success might well lead one to consider that scientific knowledge must consist of either true knowledge or is at least becoming more true over time, with the hope of even-tually reaching ultimate truth. But that simply cannot be the case. I argued in earlier chapters that we cannot *prove* that scientific laws and theories are true nor can we prove they are false because they are always underdetermined by data and also because of the unavoidable intermingling of theories that prevents one from investigating any theory in isolation from all the others. We now see that scientific evolution is also contingent on contemporary conditions and thus we cannot prove that our current theories were *destined* to eventually appear.

Thus the parallel between scientific progress and biological evolution is com-plete, and a necessary consequence is that the state of scientific knowledge that we now have is also the result of many turns taken by the scientific community in the course of its history, based on judgments that were influenced by contingent events, leading to the conclusion that it is not necessarily the *inevitable* result of scientific progress. Just as if we could run the clock again there is no guarantee that we would have the same kind of biological forms emerging, so it is with sci-ence, that the theories that emerge may look nothing like what we have now. We cannot justify the belief that we are moving toward a preordained goal, the "truth" if you will, however strong the sense of progress may be. Scientific evolu-tion, just like its biological counterpart, is not teleological.

21

The three trees of scientific knowledge

[This chapter builds on Charles Darwin's metaphor of the Tree of Life to construct two other tree metaphors that illustrate the nonteleological nature of science. One is the Tree of Scientific Paradigms that exemplifies the process by which scientific paradigms evolve, with new ones emerging over time that have resulted in their present variety and diversity. The other is the Tree of Science that represents the evolution of scientific knowledge *as a whole*.]

If the present state of scientific knowledge is, as argued in the previous chapter, but one of the many possible states that could have emerged out of all different historical paths that might have been taken, then we cannot argue that it is progressing toward "the truth," if that concept refers to something objective and unique. But if we are not moving toward truth, then where are we headed? How do we explain the fact that scientific progress is undeniably leading to greater degree of control over our society and environment? While there can be debate over whether this increased power and control over the world is for the better or worse, the fact of it is rarely in dispute. Scientific progress has put people into space, produced new drugs that promise hope for sufferers of previously fatal illnesses, and created many new things that have improved our lives immensely even though the fruits of such efforts are distributed unequally across the globe. All this has been a consequence of the increased precision and scope of our scientific theories and argues in favor of the position that the scientific community must be doing *something* right. But what is that something?

As we have seen, some have argued that the fact of science being so successful is *by itself* sufficient to claim that we must be approaching truth, and that the reasons for it will be discovered later or may prove to be ever elusive. Could it be that our inability to prove that scientific theories are true or at least heading toward truth is a temporary shortcoming due to the current state of epistemology? That is always possible but the basis for such optimism is weak. Appealing to truth as the *presumed* basis for science's success has a disturbing similarity to the arguments of early natural philosophers and current theists who suggest that the world seems to function so well that it has to have been designed and thus presupposes a designer. We know that such appeals to incredulity are risky and usually just signs of desperation. After all, many seeming mysteries of the past that once seemed to defy any rational explanation are now understood as the working out of scientific laws that were later formulated. Incredulity is not a sound basis for any view since all it might connote is lack of imagination.

There seems to be a way out of this particular conundrum that preserves the traditional beliefs of truth and scientific progress. It could be argued that the parallelism that has been drawn in the previous chapter between scientific and biological evolution is not exact and that a key difference is that for biological evolution what I have called the *soft* environment, consisting of immediate environmental factors that change over time, plays a major, even dominant, role. This soft environment consists of the atmosphere and biological features such as animals, plants, insects, and so forth, and since organisms are part of the soft environment, changing the organism also changes that environment. An ever-changing environment means that the success of a mutation depends not only on the nature of the mutation itself but also *when* it occurs, thus undermining the idea of an inevitable goal-directed process.

In the case of the evolution of scientific knowledge, however, one could argue that the situation is different. The ultimate environment that determines the success or failure of a scientific theory is not the soft one of scientific, social, political, and religious contexts which are always changing but instead is a *hard* environment that is unchanging that consists of an objective reality. Scientists sometimes use the word "nature" to connote this abstract, objective, and permanent state consisting of the principles and laws that govern the physical world, its fundamental unchanging constituents, and the theories that describe them. I will capitalize this as "Nature" to distinguish it from the "nature" that consists of the changing physical environment we live in. In this approach, scientific knowledge is believed to consist of *discoveries* about Nature and are not inventions of the human imagination, and thus research only *reveals* what already exists in Nature. Understood in this sense, nature changes but Nature does not, only our knowledge of Nature changes over time.

In this view, the environment in which scientific theories evolve is an unchanging Nature, unlike in the case of biological evolution with its changing nature. Thus, in the competition of rival theories for acceptance, the *time* of occurrence of the competition should be irrelevant. This does not mean that in the short run, incorrect theories will always lose or that other subjective factors do not play a role. Social and historical contexts always play an important role in the generation of new ideas and their acceptance or rejection. But whatever theory becomes dominant does not change Nature in the way that evolutionarily successful organisms in biological evolution change nature. In the long run, incorrect theories will eventually collapse under the weight of their own contradictions because eventually they will prove themselves to be incapable of describing the fixed, unchanging Nature. Even though we can never know if our current theories are the correct ones, we can be confident that correct theories will eventually win.

The problem is that we cannot show that such a fixed, unchanging Nature exists. What I will propose is this chapter is an alternative resolution of the paradox that accepts the current epistemological limits of science as fundamental, and yet can explain the increasing power and success of scientific knowledge. In this view there is no assumption of the existence of a fixed, objective Nature of fundamental entities and theories and laws.

My goal must necessarily be limited. Given that even in the highly rigorous realm of mathematics we cannot prove that its theorems are true, and that we cannot prove that scientific theories are true despite having overwhelming evidence in support of them, it would be presumptuous to claim that *any* theory of epistemology, with its much more ambiguous evidentiary basis, could be proven to be true. All I will be aiming for is *plausibility* and in order to do so I will heavily exploit a metaphor based on Darwin's Tree of Life.

Darwin's most famous and influential book *On the Origin of Species* contains just a single figure. It consists of a schematic representation of the divergence of taxa where time flows upward. Starting from a few points at the bottom that represent species at a very early time in the Earth's biological history, lines flow upward and proliferate as branches emerge at various points as new species emerge, until we get to the canopy of the tree that represents the current time and where we have many more lines than we started with, representing the rich biological diversity that exists today. Figure 21.1 looks more like a tree than Darwin's original figure but its meaning is the same.

All the end points on the canopy of this Tree of Life represent current organisms and species while those branches that terminate before reaching the canopy are dead organisms and extinct species. So one small region on the canopy would represent humans, while other regions would represent all the varieties of apes, chimpanzees, dogs cats, fish, reptiles, insects, plants, and so on. Dinosaurs, dodos, Javan tigers, and other extinct species would be represented by branches that end somewhere in the interior of the tree.

Note that there are gaps in the canopy between the end points and these gaps represent the differences between current species. Anti-evolution creationists often seize on these gaps, claiming that they represent "missing links" in the evolutionary record and thus disprove evolution. They seem to think that there should be a smooth variation among all *existing* species, represented by a smooth and continuous canopy for the tree. They also make specious arguments along the lines of "If humans evolved from monkeys, why are there still monkeys?" Such comments reveal serious misconceptions of the evolutionary process as being that of a *linear* progression in which the monkey species is on the same tree limb as humans but somewhere in the interior of the tree, that is, at an earlier

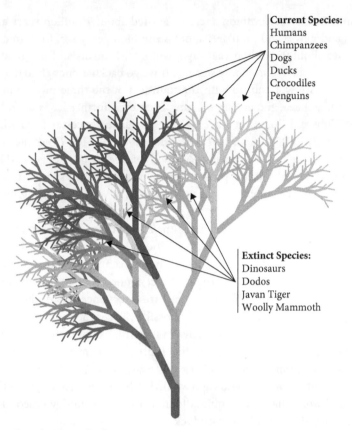

Current Species:
Humans
Chimpanzees
Dogs
Ducks
Crocodiles
Penguins

Extinct Species:
Dinosaurs
Dodos
Javan Tiger
Woolly Mammoth

Figure 21.1 The Tree of Life
Credit: Josh White

point in time. But humans did *not* evolve from monkeys. Instead we share a common ancestor, which is quite a different thing. Humans and monkeys occupy different points on the canopy and there is no *direct* connection between them along the surface of the canopy nor should we expect to see one. Those gaps do not represent "missing links."

In order to see the connection between two species, humans and chimpanzees for example, one would have to start at the point on the canopy representing any human being and trace its ancestral path by heading back down that limb into the interior of the tree until one reaches the branch point that connects to the limb that leads back up to chimpanzees in the canopy. That branch point represents the *common ancestor* of the two species. The greater the divergence between two species, the farther back in time (i.e., farther down the tree) one must go before one reaches the common ancestral branch point. So for humans and chimpanzees the common ancestor existed about 7 million years ago, for humans and new world monkeys about 45 million years ago, for humans and dogs it was about 95 million years ago, while for humans and the great white shark it was about 490 million years ago. If we go back far enough to over three billion years, we will end up with the founders of the three major kingdoms of bacteria, archaea, and eukarya. Going back even further leads us to LUCA, the Last Universal Common Ancestor, the postulated ancestor from which all life evolved (Pross 2012, 88, 90). The website *Time Tree: The Timescale of Life* at http://timetree.org is based on the book *The Timetree of Life* (Hedges and Kumar 2009) and gives results for the point of divergence of any two species. It is an excellent resource for mapping out this type of tree, as well as being highly enjoyable to explore.

Some critics of evolution even make the risible claim that if evolution were true, we should see something like a "crocoduck," a hybrid consisting of a crocodile's head and a duck's body. Such a hybrid cannot and does not exist because evolution does not require such a direct connection between any two current species. Neither crocodiles nor ducks are direct ancestors of the other nor can they interbreed so a crocoduck is impossible. To see the connection between those two species, one would again have to start from a point on the canopy that represents a crocodile, go back down that limb into the interior of the tree until one reaches the branch that goes back up to the duck. That branch point that occurred about 240 million years ago would be the common ancestor of the two species and would likely look quite different from present day crocodiles and ducks and definitely not like a crocoduck.

One of the most appealing features of Darwin's Tree of Life (at least for me) is that it shows so clearly that every point on the tree is connected to every other point. In other words, *every single one of us can, at least in principle, trace a biological link to every other organism that exists now or has ever lived,* be it human, other animal, insect, plant, or bacteria. However much we may differ superficially, deep down we are all related to one another. We are all cousins. It also makes little sense to speak of any one current species claiming to be "more evolved" than any other, since every point on the canopy has equal status since they all started at the same time. However, the idea that humans are not the apex of the evolutionary process but just one among many equals can be disturbing to those who see us as somehow special.

The biological Tree of Life is a powerful metaphor for the evolution of living organisms and a similar metaphorical tree can also be constructed (Figure 21.2) to represent the *growth of scientific knowledge* over time that I will call the *Tree of Scientific Paradigms.*

As with the Tree of Life, where one starts is somewhat arbitrary and I have chosen the bottom to represent the early days of modern science starting around

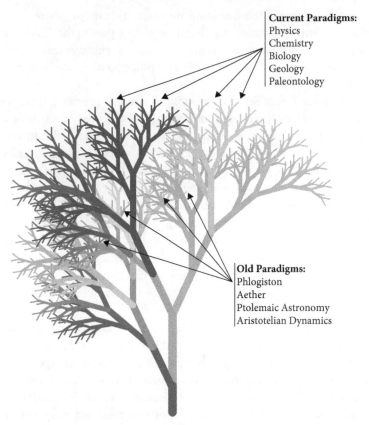

Current Paradigms:
Physics
Chemistry
Biology
Geology
Paleontology

Old Paradigms:
Phlogiston
Aether
Ptolemaic Astronomy
Aristotelian Dynamics

Figure 21.2 The Tree of Scientific Paradigms
Credit: Josh White

the sixteenth century where there was much commonality and little specialization among natural philosophers, except for the more ancient fields of astronomy, geometrical optics, and mathematics. Those were the only fields that we
could consider to be dominated by paradigms. As time went on and we go up the
tree, more paradigms emerged to meet specific needs in different areas and this
multiplication of paradigms led to increasing specialization and the formation of
new disciplines, with each discipline having its own vocabulary, structure, and
research protocols. So natural philosophy (or "science" as it became later known)
split into physics, chemistry, biology, geology, and so on, and later each of those
disciplines divided into subdisciplines such as physics breaking up into atomic,
nuclear, condensed matter, biophysics, and the rest. And the process continues
to this day as each of those subdisciplines split into yet narrower specializations
with their own paradigms. This diverging behavior can be represented by branch

points in the interior of the tree, until we reach the canopy where each point represents a currently existing subdiscipline and its paradigms. Those theories and their associated paradigms that were rejected for whatever reason and effectively ceased to exist (such as the aether and phlogiston) would be stumps that ended in the interior of this tree.

This schematic Tree of Scientific Paradigms also provides a way of explaining why the extreme reductionist goal of unifying all the areas of scientific knowledge under one umbrella comprising one set of equations is unlikely to be ever attained. This idea that is sometimes stated as "all biology is chemistry, all chemistry is physics, and all physics is particle physics" is based on the idea that the smaller the size scales of the constituent elements that form the basis of study for a discipline, the more "fundamental" that field is. But while there is a clear hierarchy of size scales of constituents, that does not imply a corresponding hierarchy of theories and concepts, as has been discussed earlier in chapter 4. Two points on the canopy of the Tree of Scientific Paradigms representing two different subdisciplines do not have a direct link connecting them, just as two species on the canopy of the biological Tree of Life do not have a direct line of connection.

Physics, chemistry, and biology are separated by largely *unbridgeable* gaps, just like humans and dogs and sharks on the biological evolutionary tree. The concepts, laws, and principles that govern subdiscipline A emerge as one goes up the limb to the point on the canopy represented by A, and similarly so for subdiscipline B. There is no reason to think that there will be a way to *directly* derive the paradigms of B from those of A or vice versa, any more than we can think of humans as descendants of monkeys or vice versa. In order to see the connections between subdisciplines A and B, one has to go back down A's limb into the interior of the Tree of Scientific Paradigms until one reaches the point in the past where the split in paradigms occurred that led to the two different points on the canopy (which is the analog of a common ancestor in the Tree of Life), and then follow that other limb back up to subdiscipline B. It is not clear what *practical* benefits one would gain by this exercise but one would undoubtedly obtain deeper insights into the historical evolution of science.

The relationship between different disciplines can, if desired, be characterized in terms of the different size scales of its constituent entities, but the theories in the sciences that deal with the larger size scales are not necessarily derivable from those of the smaller, thus making infeasible the extreme reductionist goal. They are just different places on the canopy and each will construct its own new knowledge based on where they are now and the interesting questions that the present state of knowledge in that field generates. The conceptual distance between any two disciplines may increase or decrease but whether this happens has less to do with the sizes of their constituents than with the nature of the questions that are asked in each discipline and the subsequent directions of scientific evolution created by those questions. If different disciplines evolve so that they have a commonalty of interests, as occurred with the age of the Earth, then they will pose similar questions and invent new concepts that will bring them closer together and perhaps even overlap. But this synthesis will occur because of new conceptual developments rather than by taking a reductionist approach.

The tree metaphor for the evolution of scientific knowledge can provide additional insights into the evolution of science. For as time passes, the tree grows and consequently the canopy expands and spreads and more points on it will appear and the gaps between points on the canopy can also increase. This means that as time goes by, our knowledge overall and within each discipline will increase and be seen to progress, but our ignorance of areas other than our own can increase or decrease depending on whether the distance between the disciplines increases or decreases, with the former being more likely. In addition, more and more subdisciplines will emerge.

As an aside, the word "scientist" was itself coined only as late as 1834 by academic William Whewell, who worried that the fields of chemistry, mathematics, and physics were spreading apart and that there was need for a term that would emphasize their commonalities. But its introduction was not universally accepted, especially in the United Kingdom where the older labels of "natural philosopher" and "man of science" or "savant" were preferred. The word gained much wider acceptance when the prestigious science journal *Nature* announced in 1924 that it would allow contributors to use it. The word quickly caught on, with the Royal Society of London, the British Association for the Advancement of Science, the Royal Institution, and Cambridge University Press adopting it.

Historian Edward Grant says that the use of the word "science" as a unifying concept appeared much earlier than the word "scientist."

> The medieval Latin term *scientia* was used for mathematical astronomy geometric optics, music or harmonics, and mechanics, especially statics, which was known as the "science of weights" in the Middle Ages. When these disciplines are discussed, we are justified in translating *scientia* as *science*. Contrary to commonly help opinion, the word science was not first used in the nineteenth century but was first employed, in a limited sense and in its Latin form in the late Middle Ages. (Grant 2004, 21)

Whewell's concern about the sense of unity of scientific knowledge disappearing was echoed later in the famous *Two Cultures* lecture series given by C. P. Snow, where he worried that a common foundation of education and culture was being lost and that people who were quite learned in fields such philosophy, literature, politics, economics, and sociology no longer had even the most elementary knowledge of science (Snow 1964). We see that the emergence of the two worlds that Snow deplored many decades ago (and that concerned Whewell even earlier) is not only unlikely to reverse itself, it will instead get accentuated, with many more worlds emerging as more and more disciplines become self-contained, esoteric, and separate from one another, making themselves largely unintelligible to everyone except their own practitioners, though their practical consequences will benefit many. This process is irreversible and we have to give up the hope of returning to an idealized time when it was possible for highly educated people to have reasonably in-depth knowledge of a wide area of knowledge, although it is doubtful if such a state ever existed or is just a rose-tinted view of the past.

The central questions posed in this book are why science works so well and what its relationship is with an objective and unchanging Nature or truth, assuming the latter exists. It seems like one is forced to make a choice between just two options. The first is that science works because it is true or approaching truth and the methodological problems with understanding why it is true must be due to inadequacies in the arguments of the epistemologists and philosophers of science. This is the option that is probably favored (at least implicitly) by most scientists since it *seems* plausible and it retains the appealing idea that we are seeking truth and approaching it. I have been arguing for the other choice, that those historians, sociologists, and philosophers of science who argue that there is no guarantee that science is heading toward truth are right. That puts the onus on me to explain how it can still be the case that scientific theories work so well and seem to be working better all the time.

I believe that the solution to this problem lies in looking more closely at our models of Nature and knowledge. At the risk of overworking the tree metaphor, it is possible to construct a third tree that I will call the *Tree of Science* (Figure 21.3) that can explain the fact of scientific success while conceding that it may not be taking us nearer to the truth. It must again be emphasized that such metaphors are purely suggestive and illustrative and are not susceptible to proof. They are meant to provide a heuristic way of understanding the nature of science that is as free from internal contradictions as possible and to provide a *plausible* explanation of historical development. If it achieves that limited goal, then the model is useful.

This new Tree of Science represents *scientific knowledge as a whole*, where any point on the canopy or on any limb in the interior does not represent a specific discipline or subdiscipline as was the case of the Tree of Scientific Paradigms. It instead represents *all* of science at a particular time. The paradigm revolutions that Kuhn spoke of do not signify the victory of right ideas over wrong since there is no objective way of determining that that was the case. Instead they are simply branch points in this tree. After each such decision point, the scientific community arrives at a consensus (though not unanimous) judgment

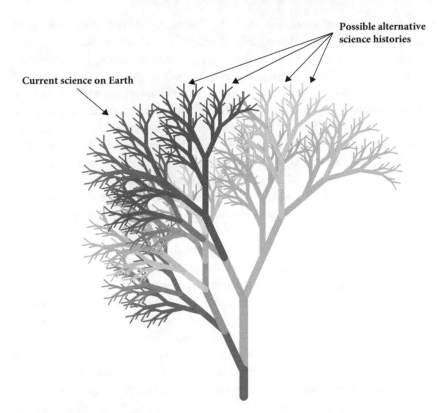

Figure 21.3 The Tree of Science
Credit: Josh White

as to which paradigm to use in future and thus *scientific knowledge as a whole* proceeds along one particular limb, where that limb now represents the state of *all* scientific knowledge, not just that of the victorious paradigm as was the case with the Tree of Scientific Paradigms. The result of decisions that are made at each branch point will result in scientific knowledge *as a whole* ending up in a particular small region of the canopy. So in this Tree of Science metaphor, only one limb, with its zigzag path, represents the actual historical path that the *entire constellation of scientific knowledge* took to get to where we are now, as it twists and turns on its way up to the canopy. Hence a single point on the canopy of Figure 21.3 represents the *entire canopy* of Figure 21.2.

What do all the other limbs and regions on the canopy of Figure 21.3 represent? All the other limbs are *hypothetical* ones that represent all the other possible paths that scientific knowledge *might have taken* if circumstances had been such that different paradigm choices had been made in the past, as discussed in the previous chapter and represented by Figure 21.2. They represent the roads *not* taken and in that sense they are *virtual* limbs that do not "physically" exist (i.e., do not represent actual histories on Earth) but need to be imagined. If, when confronted in the past with the need to make a paradigm choice, the community of scientists had chosen a different theory from the one that was actually selected (as was represented in Figure 20.1 in the previous chapter), then we would have followed a different limb and arrived at a different point on the canopy and our understanding of the laws and theories and constituents would be different from what it is now. The totality of scientific knowledge represented by that point would not be wrong but just *different* from what we have now, just as different end points on the biological Tree of Life represent different species, with no one species being the right one and all the others wrong. If we had taken one of those alternative paths, we would now be interested in different questions and our final state of knowledge, though different, might have worked just as well, or possibly even better, in enabling control of the world. (Those who are familiar with the Many Worlds Interpretation of quantum mechanics [Byrne 2007] may detect some superficial similarities with that model, with paradigm choices corresponding to the results of acts of measurement that spawn new universes, except that what is being created here are not physical universes as in the Many Worlds Interpretation but metaphorical "universes" of scientific laws.)

If one looks for them, one can see such contingency driven evolutionary trees of knowledge all around us, especially in the case of technology. For example, if one went shopping for automobiles about a hundred years ago, one had the choice of cars that were driven by gasoline, steam, or electricity. Starting a car using steam engines took a long time, about 45 minutes, to get the water to boil and generate the steam to power the car. Electric cars started quickly but at that time had a range of only about 30 miles and a top speed of 20 mph and since many houses did not have electricity, opportunities for recharging were few. All these drawbacks made electric cars impractical. As a result of these contingent factors, the gasoline-powered internal combustion engine became the preferred choice and cars since then have evolved according to that choice. The cars we have now are far more sophisticated and powerful than those early models, so much so that we find it hard to imagine alternatives to them (Arehart 2016).

Although the choice of gasoline engines might have seemed obviously better at that time, with hindsight we now see that it is not necessarily so. With the arrival of the widespread availability of electricity and increased concern about the impact of fossil fuels on the environment, conditions have shifted increasingly in favor of electric cars and we now see a resurgence of interest in them. But the electric cars that emerge now, and which may displace gasoline-powered cars completely in the future, will be based on many of the features of the gasoline powered cars that currently exist and will be quite different from the cars that we might have had if the conditions had been right for them a century ago and they had become the dominant mode of production then. The limb on the Tree of Science for current electric cars branches off at a much later point than for ones a century ago, and thus will reflect the influence of its long gasoline-powered history, just as the body shape (morphology) of humans is strongly influenced by the fact that our evolutionary branch dates back to our common ancestry with fish (Shubin 2008).

The revival of electric cars now after having been rejected earlier is a rare case of the resurrection of an old idea. As a biological parallel, we know that birds evolved from dinosaur ancestors. Could it happen that at some point in the future, dinosaurs could re-emerge from the lineage of some current bird species? While theoretically possible, that seems practically unlikely because all the favorable contingent factors that led to the emergence of dinosaurs in the past are unlikely to repeat themselves, unlike in the case of electric cars. We are tightly constrained by our heritage in all areas, be they biology, science, or technology, and reversing the process or starting completely anew is highly unlikely.

There are two immediate consequences that emerge from this Tree of Science metaphor for the evolution of scientific knowledge as a whole. The first is that because the tree never stops growing, our knowledge will always increase as time increases. There is no limit. It also means that we are not growing *toward* some fixed point that represents Nature or "the truth" or "the real world" where our journey will end, just as biological evolution is not heading toward some state of perfection where the process ceases. All we can say about scientific knowledge is that like its biological counterpart, it is something that evolves *from* the past, not *toward* a specific future. Knowledge is not something fixed and external (i.e., the "truth") that we slowly unpeel and recognize. Instead, all scientific knowledge is something that is *constructed by acts of human creativity* and increases with time. Hence what we can and will know in the future depends on what we know now, because it is what we know now that determines what questions we ask and that guides the search for the answers, which in turn creates new knowledge.

It is worthwhile to re-emphasize a key point that follows from this model. If at some crisis point in the past, the scientific community had chosen a different paradigm than the one it historically chose (and would thus, according to our retrospective whiggish understanding of history, be perceived as having made an "error"), all that would have happened is that we would have followed a different path to the canopy and occupied a different region of space on it. But having arrived there, *we would see that state as being as correct and inevitable as we now see our current state.* Thus science will always appear to make "progress," acquiring new knowledge that is considered better than that which it replaced, while at the same time not moving inexorably toward a preordained goal, *because there is no preordained goal.* We can thus simultaneously see science as evolving and describe the process as progress while denying that it is approaching "the truth," as long as the latter concept refers to one particular point on or beyond the current canopy of the Tree of Science.

In the model that I propose here, there is no end point to the search for knowledge. The potential for new knowledge is expanding even as our knowledge increases. We will never "know everything" because there is no "everything" to know. New knowledge can and will always be constructed out of existing knowledge. As long as the word "truth" is used to signify a *unique* state of knowledge to the exclusion of all else, we cannot prove that it exists or that we are moving toward it. There is no justification for thinking that our present state of scientific knowledge, which is represented by the region of space we occupy on the canopy of the Tree of Science, is "the truth" or that there exists a *unique* truth at all. Even if there were, how would we know if we had reached it? As stated earlier, scientific research is not like a video game where bells ring and screens flash to indicate that one has reached one's goal. We are always in a state of doubt as to whether we have got it right. What we believe to be true is always provisional and the result of a consensus judgment by the community of scientists at any given point in time. There are always new questions that can be asked and new avenues to be explored. Indeed, this is what makes science so endlessly fascinating.

The famous concluding passage from Darwin's classic work, in which he refers to the never-ending process of biological evolution, applies equally well to the endless process of knowledge evolution.

> There is grandeur in this view of life, with its several powers, having been originally breathed into a few forms or into one; and that, whilst this planet has gone cycling on according to the fixed law of gravity, from so simple a beginning endless forms most beautiful and most wonderful have been, and are being, evolved. (Darwin 1859, 490)

22

Resolving the Great Paradox

[This chapter uses the Tree of Science to resolve the paradox of how scientific theories can work so well and be so successful in explaining the world around us despite the lack of any assurance that those theories are true or that they are even approaching something that we can call the truth. It argues that the ideas of truth and correspondence with reality are unnecessary for understanding the success of science and are actually a hindrance and a distraction because they raise metaphysical questions that cannot be resolved.]

The Tree of Science metaphor explains why science need not be progressing toward the truth but still leaves us with the remaining paradox of how science has enabled us to make such tremendous advances in technology. This puzzle arises because of the long-standing assumption that the ability to *predict* and *control* things is identical with the level of truth of the theories being used to achieve that control. The fact that we seem to be steadily achieving increasing technological success and control over our environment is believed to be due to the fact that we are getting closer to "the truth" about the way things are. Some suggest that the success of science would be miraculous if that were not the case.

But why do we believe truth and control are so intimately connected? If we disentangle those two concepts, then that puzzle goes away. Our tremendous technological success is simply a measure of how well we can control things, especially those that are inanimate. But there is nothing that forces us to believe that control over nature can be achieved *in only one way* and that increasing control implies increasing truth. That would be similar to thinking that there can only be one type of automotive engine or one form of airplane because the ones we have now work so well. There may *seem* to be just one possible form at any given time, but as we have seen, that may be simply due to the constraints imposed on our imaginations by the historical path we have taken.

Using the tree metaphor again, control is like the ability to shake the Tree of Science so that its metaphorical "fruits" (i.e., the concrete results that have enabled us to make such immense progress) fall down so that we can enjoy them. That part of the canopy we presently occupy represents our current state of knowledge and is what we can grasp now to shake the tree, and the fruits on the limb leading to that point on the canopy represent the technology spawned over time by that state of knowledge. It is the use of the *logic of science* outlined in this book that enables us to exercise such control by enabling us to make reasoned judgments to arrive at scientific theories that work well and avoid sterile speculations of things that are neither naturalistic nor testable.

But we could have shaken the tree by grasping other points as well, each representing alternative states of knowledge that would have been reached if science had taken different historical paths. Grasping different points on the canopy will result in us shaking different limbs and the fruits that fall will largely be those on those limbs. But those limbs are *hypothetical* ones, representing the paths that scientific evolution *might* have taken but did not due to contingent factors. Would those hypothetical fruits be the same as the ones we currently have, suggesting that whatever path science took, we would have ended up with the same concrete results? We don't know for sure but it seems unlikely. If the fruits are different, we cannot be sure that what we have now are the best we could have had.

Biological evolution is based on what is the best option for *short-term* reproductive success of an organism in the immediate environment. It is not based on seeing what might be best in the long-term future. Similarly, science evolves by determining what paradigms work best at particular times in history. The fact that where we ended up on the Tree of Science may not be the best of all possible outcomes does not imply that our predecessors made errors in making the paradigm choices they did. Each choice was made on the basis of scientific logic and was rational at the time it was made. It is simply that the choice was not made on completely objective grounds but also depended upon contingent factors that existed at that time. A different set of circumstances, a slight change in the temporal spacing of events, and we could have ended up with a very different scientific understanding of how the world works from what we have now.

Other places on the canopy of the Tree of Science representing other "truths" may be even better for the purposes of control, at least in some areas, but due to the paradigm choices made in the past, we did not arrive there. We might have had a different level of control and different results if we had taken different paths and we cannot know how those would compare with ours. Some other parts of the canopy may well have given us better results. Just because we occupy and are aware of just one region and our theories seem to work pretty well, we should not fall into the trap of having a Panglossian sense of hubris that where we are now is the best of all possible worlds. The writer of the biblical Book of Ecclesiastes captured well how, when it comes to determining outcomes, contingent factors can overwhelm more objective ones.

> I returned, and saw under the sun, that the race is not to the swift, nor the battle to the strong, neither yet bread to the wise, nor yet riches to men of understanding, nor yet favour to men of skill; but time and chance happeneth to them all. (King James Bible, chapter 9, verse 11)

Are there *any* universal scientific truths at all that are independent of our choice of paradigms? In other words, whatever choices we might have made in the past, and wherever we ended up on the canopy of Tree of Science, are there some specific results that would always have emerged? I mentioned earlier the scientist who said that water being comprised of two hydrogen molecules and one oxygen molecule had to be a rock-solid fact. He said that he could not possibly conceive of how it could ever be shown to be anything else. Surely the existence of the electron is also such a fact? Surely that the speed of light is independent of the state of motion of the source or the observer is a similar fact? It does not seem, at least on the surface, that those things could possibly be false *without rejecting the entire scientific framework that we now have.*

And indeed that is the case. To conceive of an alternative framework that does not contain these "facts," one would have to go back down our particular limb of knowledge on the Tree of Science to a branch point and then go back up a different limb. How far we go back down will influence how different the final state would be from what we have now. But while the known historical development of our current science can serve as a map for our way down, that upward journey on a different limb requires *constructing an entire alternative history of scientific progress out of whole cloth*, and doing so is immensely difficult, I would say impossible, because of the vast number of variables at play. To try to conceive of an entirely different scientific framework based on a different history from the one we actually have boggles the mind. The immense difficulty of doing so prevents us from even attempting it and misleads us into thinking that what we have now must be unique, the inevitable outcome that was destined to happen.

One sees this in the case of biological evolution. For a long time, people believed that human beings were the pinnacle of life, the ultimate act of creation in the image of God. Even among many of those who accepted that humans were part of the evolutionary process, it was believed that it was *intended* that we would eventually emerge. Anything else seemed inconceivable. What was hard to accept about Darwin's mechanism of natural selection was that we may be just one of the many possible branches that emerged out of the evolutionary tree, that happened to survive due to a particular set of soft environmental circumstances at each point along the way. What is worse, we may not even be the best of all possibilities, let alone the best of all the currently existing species. One can conceive that there could have been a superior life form to human beings that did not come into existence because the evolutionary branch that would have resulted in it diverged from the one that led to humans, and this divergence occurred at a time that was not the most propitious for it. Hence that branch died out, even though its end result might have been an alternative species better suited for the present time. This realization can be too humbling a comedown for many people to accept. To go from the pinnacle of creation to just an accidental byproduct of the evolutionary chain is a very steep drop. It should not be surprising that Darwin's ideas were fought with such vehemence, especially by those who had been conditioned to thinking of humans as very special indeed.

It is possible to speculate on other reasons why Darwin's ideas were hard to accept for many people. Part of the reason may be simply lack of imagination and the impossibility of doing experiments to test alternative models. The human body is such an extraordinarily complex and finely tuned instrument that it is hard to imagine how it could be anything else. But it is possible to ask some speculative questions. Why do we have ten fingers and toes? Would more or less have been better for life as we know it now? Could there have been a branch along the evolutionary path in which we could have evolved with eyes that could see behind us as well or at least provide a wider range of view? Or perhaps have less susceptibility to cancer and heart disease? Having an extra pair of hands might have been helpful, especially now given the demands of modern life and the ubiquity of mobile phones and texting. After all, there are other life forms that have some of those features. There are an infinite number of such questions that can be posed but they are impossible to answer in any concrete way. Trying to imagine all the other possible ways that humans might have evolved is a mind-boggling exercise. So it is convenient to just accept the way things are and simply try to explain how they got to be that way. But we often forget that that is a deliberately chosen limited goal that we have adopted out of necessity, and slip into thinking that because we can plausibly explain how things got to be the way they are now, that must be the *only* way they could possibly be.

The same criticism can be leveled at our present perception of the state of scientific knowledge. Given the unquestioned success that this knowledge has had in achieving technological success, and the intricacy and sophistication of our present scientific theories, it is difficult to even contemplate what might constitute a better theoretical framework. But just as in the case of biological evolution, there may well have been branches on the Tree of Scientific Paradigms that led to other points on the canopy of the Tree of Science that would have provided us with theories of even greater sophistication, complexity, beauty, and utility, but those branch points occurred at times when the theories that would have led to these particular end points did not have much chance of success. Consequently, we ended up where we are now. It is always possible that future scientific developments could result in a path that takes us back to one of those better states of knowledge but there is no guarantee that it will. What we should be cautious about is falling into the trap of believing that there can be nothing that is better than what exists now or that anything better can only be derived from whatever we have now.

There is one group of people that does specialize in imagining possible alternative worlds and that is science fiction writers. They are the ones who try to break free from what currently is and try to envisage what might be. But even they tend to limit themselves. Some of the most respected science fiction writers are those who largely accept the constraints imposed by the present state of scientific theories but loosen them slightly here and there, such as to allow human beings to reach other planets in other galaxies or to have extraterrestrial beings come here. Trying to envisage an entirely different scientific knowledge structure would not only be difficult, the result might be too strange for audiences to accept.

For example, it has been suggested that there are five necessary conditions for life as we know it to evolve: (1) a temperature range that allows for liquid water; (2) water; (3) energy sources; (4) nitrogen (to form amino acids); and (5) oxygen (Lemonick 2014; McKay 2014). That list is derived from studying life on Earth. But are the laws of biology that we have discovered on Earth *universal* laws? Could they be in fact contingent laws that created life because of the conditions we have here? Could we have life that is not carbon-based as is all life on Earth but based on other elements, with silicon replacing carbon and chlorine or fluorine replacing oxygen (Catling 2013, 58)? Could there be life forms that do not need water? Trying to envisage this requires the creation of an entirely novel world that is untethered to ours, and doing so except in the vaguest and most general terms is not easy. But at least with biology, since we have many different forms of biological life even here on Earth, and we understand the evolutionary branches that brought them into being, we can conceive of some alternative forms, such as intelligent beings that look less like humans and more like insects or reptiles. We also have had regions of the Earth that were separated from the rest and thus enabled different life forms to evolve independently, such as the kangaroos in Australia. Similarly, in the case of language evolution, we have many different examples of independent evolution in the form of indigenous languages around the world, though many of them are going extinct.

But in the case of the Tree of Science, we have no similar points of comparison. The science that emerged from the time of Galileo and Newton has been like a phenomenally successful invasive species, permeating into every part of the globe and overwhelming any local forms, thus obliterating any possible alternative scientific structures that might have emerged in an isolated society. Science has become global, with scientists everywhere using the same paradigms in very similar ways. There is no scientific equivalent of an isolated island or continent. Thus we have experience with only one outcome and that is the world of science we have now that occupies a tiny part of the canopy of the Tree of Science. The only way that we can experience possible alternatives untainted by our own history is if we could communicate with extraterrestrial beings whose evolution of scientific knowledge occurred completely independently of ours, along a different branch of the Tree of Science.

In many science fiction stories of encounters with extraterrestrials, the climax is often the meeting with them. The tension is often based around biological and social questions of what they look like and whether they are friendly or hostile. But one could ask deeper questions. Were I to encounter an extraterrestrial, what I would be most interested in are the laws of science they used to develop the advanced technology they must possess in order to have made the journey to us. Do they use Newton's laws? How about the theories of relativity? Quantum mechanics? If they have different laws, are they such that we can see a connection between the two systems, suggesting that the laws are unique and that the differences are small and superficial and can be explained by some fairly straightforward transformations between them? Or are they so completely different as to be unrecognizable, suggesting that the evolutionary path of their Tree of Science took a completely different turn from ours?

In other words, if we were to encounter extraterrestrial beings whose scientific knowledge evolution had been independent from our own, would there be any conclusions at all that we would have in common, which would suggest that there are indeed at least some things that have an objective existence? While it may seem plausible that more complex theories such as quantum mechanics and relativity (whose fundamental features are considerably removed from everyday experience) could be human constructions that have no universal validity, there are other laws that are so simple and easy to understand that it is hard for us to imagine how they could not be universal. One such case is the law that states that the speed of light in a vacuum is the same for all observers and sources irrespective of how they are moving. The concept of speed is so intuitive, and its measurement so direct, that it is hard to imagine an alternative framework. But while it may be tempting to think that such a law must be universal, the lack of viable alternatives to it may simply be due to our lack of imagination.

A science fiction short story by Ted Chiang titled *Story of Your Life* that was published in 1998 formed the basis of the 2016 film *Arrival* (Chiang 2016). It describes an encounter with aliens who arrive on the Earth on highly sophisticated spaceships. They have seven limbs and the story centers on how we learn to communicate with these heptapods. After a long time, a breakthrough occurs when it is discovered that in the heptapod world of science and mathematics, Fermat's Principle of Least Time and other forms of the calculus of variations are considered basic and simple and introductory concepts, unlike in our system where they are encountered much later in one's education and only by those specializing in advanced mathematics and physics. This gives the heptapods a different understanding of time than we have, enabling them to navigate through time in ways that are closed to us because of our ignorance of their knowledge structures. But it could also be that in the event of such an encounter, the alien scientific framework is so different that it may not be possible to even find any such correspondence and so there might be no way to tell if their science is fundamentally, not just superficially, different from ours. In their science, they may not have had the need to invent the concepts of electrons and atoms such as oxygen and hydrogen, and so their model of water may not consist of one oxygen molecule and two hydrogen molecules.

But all these scenarios, even if suggestive, are purely speculative. In the absence of an actual encounter with extraterrestrials that would give us this kind of crucial information, the Tree of Science suggests that there is no reason to think that there is a unique truth out there and many reasons to doubt it. In this way of looking at knowledge, there will never be a "theory of everything" and there is no reason to believe that we are inching closer to "the ultimate truth" in any way. In the search for knowledge, we have embarked on a quest for which there is no end. We will continue to increase knowledge indefinitely, some of it beneficial, some not so, but the journey will continue forever. This may seem a dismaying prospect to some. We like to have all our actions directed toward a specific goal, preferably one that is noble. Even if an ultimate truth does not exist or we cannot attain it, the search for truth is like the Holy Grail, a powerful motivator for many who strive hard to attain it.

One significant consequence of the conclusion that I advocate here, that there can be many different "truths," is the implication for the existence of an "objective reality." If we can have many different scientific knowledge structures, does this mean that they are all merely different representations of a single objective reality or does it mean that there is no objective reality at all? Could it be that there may be no timeless world of fundamental constituents and laws that govern their behavior that exist apart from, and independently of, our understanding of them? Physicist David Bohm said that it is possible to believe in an objective reality while simultaneously recognizing that our theories will never be able to provide a complete and final description of it.

> In this point of view, we assume that the world as a whole is objectively real, and that, as far as we know, it has a precisely describable and analysable structure of unlimited complexity. Thus structure must be understood with the aid of a series of progressively more fundamental, more extensive, and more accurate concepts, which series will furnish, so to speak, a better and better set of views of the infinite structure of objective reality. We should, however, never expect to obtain a complete theory of this structure, because there are almost certainly more elements in it that we can possibly be aware of at any particular stage of scientific development. However, any specified element can in principle ultimately be discovered, but never all of them.
>
> The point of view described above evidently implies that no theory, or feature of any theory, should ever be regarded as absolute and final. (Bohm 1957, 100)

Hence it is unlikely that there is a unique set of fundamental particles and laws such as those sought by particle theorists. As long as particle physicists are interested in this type of question, they will proceed to reveal layer after layer of knowledge, until at some point interest in this type of question, research funding to investigate it, or both, wane.

Bohm assumes that even though we can never achieve a complete and final description of the underlying reality, our theories are approaching it by getting "more accurate" and are thus getting better in some objective sense. In the model proposed in the present book, however, I have gone further and argued that even that more modest claim is doubtful. Our theories undoubtedly provide us with great benefits by increasing our control over Nature but we cannot infer from this that they are approaching truth.

John Bell, the theoretical physicist whose work on the foundations of quantum mechanics led to the remarkable theorem (Bell 1964) that is now named after him, struggled to make sense of its undermining of the belief in an objective reality and truth. He quotes pragmatic philosophers of science who say that it may not be quantum mechanics that is peculiar but that "rather the concepts of 'real' or 'complete truth' are generally mirages. The only legitimate notion of truth is 'what works.' And quantum mechanics certainly 'works'" (Bell 1981, 124). They were largely echoing Einstein's concept of reality and truth (quoted earlier in chapter 2) when the latter said that "real" is an empty, meaningless category.

As Beller describes it:

> On other occasions Einstein also rejected the notion that the "truth" of a theory lay in its "correspondence with reality." "He [the scientist] will never be able to compare his picture with real mechanism, and he cannot even imagine the possibility of the meaning of such a comparison." (Beller 1999, 155n)

The highly influential mathematician David Hilbert, whose work strongly influenced theoretical physics, had similar views about the limitations on scientific knowledge that were described as "Consistency, or rather the avoidance of a paradox, is ultimately the only safeguard in the pursuit of knowledge. We would have liked to do better for sure, but we have very little choice in the matter" (Beller 1999, 205).

The early days of quantum mechanics provide a good example of how awareness of the diversity of knowledge and interpretations can disappear over time. In the beginning, there was no understanding of the foundational aspects of the theory as to what it all meant. Indeed, there was no real theory as such. What physicists developed were computational techniques that enabled them to calculate things that resulted in agreement with certain experimental data. Quantum mechanics "worked." Whether matter was particle-like or wave-like was a topic of hot dispute for which there was no agreement, as was the question of whether the correct description should use the mathematical formalism of matrices or differential equations. Scientists used ideas from both areas opportunistically as needed to get agreement with experimental results.

But as was discussed in chapter 19, over time what is now the "official" interpretation of quantum mechanics has dominated the field while alternative views have faded from view. But this "victory" of one side was due less to the fact that the Copenhagen approach was decidedly better than to the fact that its advocates were more influential in the scientific community (Beller 1999, 39). Furthermore, illustrating the phenomenon that history is written by the winners, most of the reminiscences of that era about this particular contest were written by the "victors" and hardly any by the so-called losers (such as Schrödinger and Einstein), cementing in the minds of the current generation of physicists that the issues have been settled when that is not the case (Beller 1999, 10).

To this day, some of those disputes remain unresolved and Bell was instrumental in keeping this issue alive. While acknowledging that the pragmatic position that was driven by what works had led to tremendous progress, he also allowed that belief in reality and truth can play positive roles. "However, the notion of the 'real' truth, as distinct from a truth that is presently good enough for us, has also played a positive role in the history of science . . . I can well imagine a future phase in which this happens again, in which the world becomes more intelligible to human beings, even to theoretical physicists, when they do not imagine themselves to be at the centre of it" (Bell 1981, 125).

Bell suggests that scientific theories are similar to historical fiction.

They are like literary fiction in that they are free inventions of the human mind. In theoretical physics sometimes the inventor knows from the beginning that the work is fiction, for example when it deals with a simplified world in which space has only one or two dimensions instead of three. More often it is not known till later, when the hypothesis has proved wrong, that fiction is involved. When being serious, when not exploring deliberately simplified models, the theoretical physicist differs from the novelist in thinking that maybe the story might be true. Perhaps there is some analogy with the historical novelist. If the action is put in the year 1327, the Pope must be located in Avignon, not Rome. The serious theories of theoretical physicists must not contradict experimental facts. If thoughts are put into the mind of Pope John XXII, then they must be reasonably consistent with what is known of his words and actions. When we invent worlds in physics we would have them to be mathematically consistent continuations of the visible world into the invisible . . . even when it is beyond human capability to decide which, if any, of those worlds is the true one. (Bell 1986, 194–95)

Bell's suggestion that scientific theories are akin to historical fiction is, in my view, an extremely apt comparison but will not be congenial to those who would like to see it as closer to objective truth. Accepting it does not necessarily imply that we have to give up the idea that there is a material objective world. It just means denying that we can arrive at a unique representation of it that would constitute the "truth." It means that we have to accept that when we leave the visible world and enter the invisible that lies beyond our senses, we have to infer things for which we have no direct sensory experience, and that any knowledge we gain necessarily involves acts of human imagination, "theories" if you will, that we cannot prove to be true, just as we cannot be sure of the motives and thoughts behind the actions of historical figures or even the words they spoke. The best that we can hope for is finding out what works and what this book has tried to show is that the evidence-based, logical system that has been developed by the scientific community over time and that uses methodological naturalism, predictability, and testability, has enabled us to produce knowledge that works exceedingly well and continues to work even better over time, in the sense that they continue to solve important problems and give us increased control over our environment.

Decoupling the idea of "truth" from "control" in science is liberating because "truth" is not particularly useful when it comes to deciding between propositions. People can make truth claims for all manner of metaphysical propositions and judging the merits of those claims is next to impossible. Shifting the discussion to what works can end such sterile discussions because *to say that something works or works better than something else is to make a manifestly empirical claim that immediately demands a preponderance of supporting evidence.* In the absence of such evidence, one need not take any such claim seriously. Those who would seek to undermine the credibility of science or the weight of the scientific consensus in some area by saying that we have not "proved" it to be true would then be making an irrelevant argument. The failure of science to solve problems in this or that area cannot be used as evidence of science's weakness. Unsolved problems will *always* be present and their existence proves nothing. It is also not enough to merely suggest alternative explanations for some phenomena that science has not yet adequately addressed. Such explanations are extremely cheap to manufacture, are never unique, and thus worthless. Recall that the Raelian theory can "explain" *every* feature of the living world on Earth.

What is necessary is that any new system must be *consistent* with other areas of scientific knowledge, make *predictions* that can be tested, and that the results of the tests provide a *preponderance of evidence* sufficient to enable it to become the new accepted consensus. That, along with the ability to produce things that are useful, is the yardstick for measuring what "works" and is something that can be adjudicated by credible experts. Those who seek to undermine the credibility of science in order to advance their agendas using alternative theories have been

unable to meet that standard. Their theories are not only not true, they also either do not work at all or their performance is abysmal. Hence their claims to be
better than the scientific consensus can be dismissed.

The critics of science need to produce an alternative system *that works better
than the science we have now*. This is an extremely high hurdle for any entirely
new scientific structure to clear because of the constraint that history imposes on
the range of scientific possibilities that can emerge now. It can perhaps only be
achieved if scientific development takes place completely independently of our
own history, such as by extraterrestrials on some other planet. We cannot rule
out such an alternative structure, but it is unlikely that we will ever find out if one
exists. For all *practical* purposes, though not in principle, the science we have
now can be considered to be the best knowledge available to us.

In the search for what distinguishes science from other forms of knowledge,
we are in some sense partially back with Aristotle when he said that what makes
science so successful is its incorporation of both know-how and know-why
knowledge. But when Aristotle went on to associate scientific knowledge with
truth, that extra step was not only unnecessary, it also sent the discussion along
a road that led to an epistemological morass by entangling it with all manner of
metaphysical speculations. The strength of science is that *its know-why knowledge, in the form of its theories, directly feeds into know-how knowledge* and vice
versa. While know-how knowledge alone, in the form of technology, can achieve
great success and can spur scientific investigations, it is scientific theories and
laws that continually increase the scope and power of know-how knowledge and
enables us to make advances. As long as that combination predicts things that
can be tested and thus generate fresh bodies of evidence, and enables science to
expand and grow and be more successful, it can do so without concerning itself
with whether the know-why knowledge is true or not.

Accepting the existence of an objective, material world does not require us to
accept that there must exist some unique set of theories that represent that world
that we can label as "the truth." The best that we can hope for is not proving the
truth or falsity of our scientific theories but finding out what works. The lack
of truth as a goal, or even as one criterion to be used in the search for scientific
knowledge, does not make the scientific enterprise invalid. As long as the end result is a fairly consistent whole, as long as it is free from obvious contradictions,
and "works," then that is sufficient. To exploit the analogy with evolution yet
again, that theory works very well in explaining the diversity of organisms that
currently exist. But the validity of the theory is not dependent on the present situation being a unique outcome of the theory or the best of all possibilities. The
theory of evolution would still be valid even if a completely different spectrum
of organisms had happened to emerge from the primordial conditions. Similarly,
the fact that our current scientific theories may not be true or unique or represent

the best correspondence with reality need not be in contradiction with the fact that they work so well and are far superior to the alternatives proposed by those who seek to undermine science.

Abandoning notions of truth and correspondence with reality will be hard for some because they are so used to thinking of validating knowledge in those terms, and they may well wonder whether we really need to do so. Rather than asking whether we need to get rid of them, it is better to ask what they *add* that makes them desirable for inclusion in the first place, other than providing a warm feeling that we are on some noble quest. As Kuhn says:

> Can we not account for both science's existence and its success in terms of evolution from the community's state of knowledge at any given time? Does it really help to imagine that there is some one full, objective, true account of nature and that the proper measure of scientific achievement is the extent to which it brings us closer to that ultimate goal? (Kuhn 1970, 171)

It should be enough that our scientific theories work so astoundingly well, show every indication of working better in the future, and that the alternatives proposed to them come nowhere close to replicating their successes. What added benefit is gained to saying that the theories must be true or approaching truth or achieving greater correspondence with reality? By doing so, rather than strengthening the case for science, we actually weaken it because we cannot justify such claims to its critics. All we are doing is opening the door to endless metaphysical speculations and to pointless comparisons with metaphysical systems, whereas the claim that science works well is an empirical claim that is clearly demonstrable and competing systems of knowledge do not come even close to achieving science's level of success. *Truth and correspondence with reality are unnecessary as explanatory concepts* in science and, just like we did with the aether and phlogiston and all the other concepts that lacked any positive evidence in their favor and were also found to be unnecessary as explanatory concepts, we can regard them as irrelevant and can comfortably dispense with them as no longer serving any useful purpose.

In the continuing search for what works, we may gain a sense that we are obtaining deeper and more profound insights into the nature of our universe and that can be a powerful motivator. As long as we do not delude ourselves into thinking that we are learning *the* truth about the universe or that we are approaching any kind of finality about our scientific theories, that should be enough.

Supplementary Material

A. Bayesian methods (chapter 10(a))

The basic idea that lies behind the use of what are known as Bayesian statistical methods is that if we can assign an a priori numerical probability to a proposition being true, additional evidence can be used to modify that probability accordingly. Confirming evidence will increase the probability and contradictory evidence will lower the probability. In the opening passage of their book *Scientific Reasoning: The Bayesian Approach*, philosophers Colin Howson and Peter Urbach explain their mission:

> According to the Bayesian view, scientific and indeed much of everyday reasoning is conducted in probabilistic terms. In other words, when evaluating an uncertain claim, one does so by calculating the probability of the claim in the light of given information. Precisely how this is done and why it is reasonable is the topic of this book. . . . The rest of the book will show in detail why the Bayesian approach is the only one capable of representing faithfully the basic principles of scientific reasoning. (Howson and Urbach 1993, 1–2)

The catch is that the method does not work if what is known as the *prior probability* of a theory being true (a number that one must have to start the Bayesian analysis) is zero because then no amount of confirming evidence can raise the probability above zero. Karl Popper and Rudolf Carnap showed that this prior probability is indeed zero and hence the Bayesian approach is doomed to failure (Popper 1968a, 257; Murzi n.d.). Howson and Urbach are of course aware of this problem but argue that it is possible to make a subjective or "personalist" assignment of Bayesian prior probabilities that is not zero, and this enables them to move on and argue that the cumulative weight of evidence can be used to obtain an estimate of the likelihood of scientific theories being true (Howson and Urbach 1993, 391–92).

Andrew Gelman, another Bayesian statistician, disagrees. He says that Bayesian ideas can be applied in limited settings where the possible outcomes are well defined. But as I have pointed out, the range and number of theoretical predictions for any scientific theory are neither finite nor well defined. Gelman argues that Bayesian methods can be used to test the predicted *outcomes* of scientific theories and models in specific situations, but that it is not generally useful to try to assign probabilities of truth to the theories and models themselves (Gelman 2011; Gelman and Shalizi 2013).

B. Why scientific theories are always underdetermined by evidence (chapter 10(d))

Here is a simple, purely mathematical puzzle to illustrate an important insight about science, which is that no amount of data can uniquely determine a theory. Suppose you are given the numbers 1, 2, 4, and 8. What is the next number in the sequence?

These kinds of puzzles, where one is given a sequence of numbers and asked to predict the subsequent ones, are common and often appear on tests of logic and IQ. Most people would likely give the answer as 16. But is that necessarily the case?

Since we were initially just given a set of numbers with no explanation of their origins, for all we know, they could have been produced by a random number generator and so the next numbers in the series could also be random. Hence there could be an infinite number of solutions. But of course, there are implicit rules that are understood by puzzle-makers and puzzle-solvers alike that govern such exercises and going outside those rules is frowned upon. The main one is that the given set of numbers was not randomly generated but produced by a formula or a rule or an algorithm, and the subsequent numbers should also be products of the same method. A secondary expectation is that the predicted numbers be of the same *type* as the ones given, so in this case one would expect them to be integers and likely positive ones. So the puzzle is really asking us to find the formula (or rule or algorithm).

If one is told that there does exist an underlying pattern and that this set was produced using a formula, then one can reasonably infer that the numbers given are obtained by starting with 1 and doubling each term to get the next one. The fifth term would then be predicted to be 16, the sixth to be 32, and so on. The general formula for producing the nth term in the series would be 2^{n-1}. Such an answer would usually be deemed correct in standard logic tests.

While this is indeed one possible solution, is it the only one? The way the problem is stated presupposes that there is just one solution but in fact, there are an *infinite number of formulas* that can be generated that would produce the original series of numbers and then differ in the subsequent ones. Take the formula $2^{n-1}+(n-1)(n-2)(n-3)(n-4)m$ for the nth term, where m is an integer *that can take any value at all.* This formula also predicts correctly the first four terms but for the fifth it predicts $16 + 24m$. If we choose m = 0 we get the earlier answer of 16 but there is nothing that *forces* us to make that choice. If m = 1, then the fifth term becomes 40. If m = 2, then the fifth term is 64, and so on. Since there are infinite possible values of m, we have an infinite number of formulas that work as well as 2^{n-1} in predicting our starting data set.

But suppose we enlarge the starting data set to be explained and add a fifth term so that the starting series is now 1, 2, 4, 8, and 16. This also satisfies the formula 2^{n-1} but the alternative formula $2^{n-1}+(n-1)(n-2)(n-3)(n-4)m$ that satisfied the earlier set of starting values now only works if m = 0 and thus also gives us 2^{n-1}. So has this new piece of data enabled us to arrive at a unique formula? The answer is still no because now there is a *different* alternative formula $2^{n-1}+(n-1)(n-2)(n-3)(n-4)(n-5)m$ (where again m can be any number) that satisfies the same enlarged starting data set but predicts a different sixth term.

In general, however large we make the given initial sequence, there will always be an infinite number of formulas that can be constructed to fit it and there is no compelling reason to pick one. Of course, one could invoke Occam's Razor or some other principle of economy or elegance or parsimony in favor of one formula, but those are criteria that are not forced on one and thus may not command universal agreement. In fact, it is not just an infinite number of *values* of the number m that is the problem. I could if I wished replace the number m with an infinite number of *formulas.*

It is possible to add on sufficiently strict restrictions on the range of possible formulas or algorithmic solutions so that a unique solution is forced upon us. For example, in the sequence problem given here, by imposing the condition that the solution must only be of the form p^q, where p and q are integers, then the starting sequence of 1, 2, 4, 8 does indeed uniquely predict that p = 2 and q = n−1. But such additional restrictions are specifically designed to guarantee unique solutions and have to be externally imposed.

The analogy with science is close to exact. However much data and evidence one accumulates, one can never use them to arrive at a unique theory unless one imposes additional strong extra restrictions not based on objective factors that drastically narrow the choices. One could artificially restrict the range of competing theories by requiring that they only be of a certain type or lie within a narrow range, usually resulting in just two of them being plausible, and then use data to eliminate one. This is the model usually presented in science textbooks about how one theory emerges as the dominant one. What is obscured is that such situations are the result of a long history of scientific judgments that are hidden by the mists of time that have resulted in the limited choice.

C. Gödel's theorems (chapter 13)

To arrive at true statements in mathematics depends upon using axioms that are true and using valid rules of logic to prove the theorems that constitute the true statements. If the rules of logic that are applicable in a mathematical system are simple and transparent enough, there is usually little disagreement about their validity. Thus a true set of axioms will guarantee a consistent system of theorems and vice versa, in that we are guaranteed that no two theorems will ever contradict each other. Hence we can at least solve the problem of consistency if we can establish the truth of the axioms, though the completeness problem of whether we can prove all the possible true statements still remains open.

How do we establish the truth of the axioms that purportedly govern the properties of a system? If the system we are dealing with consists of a *finite* number of entities, we may be able to prove the axioms to be true by seeing if every single one of the entities in the system satisfy the axioms, by exhaustively applying all the axioms to all the entities and seeing if they hold true in every case. Even if the axioms do not directly relate to a set of objects, we may be able to construct a *model system* of objects in which the elements of the model correspond to the elements in the axioms and repeat this process. So, for example, we can take the axioms involving points and lines and so forth in Euclidean geometry (which are abstractions that have purely mathematical relationships with one another) and build a model system of real objects (such as points and lines that can be drawn in physical space) and see if the axioms apply to the properties of such objects. Similarly, while numbers are abstractions, we can see if the abstract rules for adding numbers in the formal abstract system correspond to what we get if we add up corresponding numbers of physical objects like chairs.

The catch is that for most systems of interest (such as points and lines in geometry and the integers in number theory), the number of elements in the system is *infinite* and it is not possible to exhaustively check if (for example) every point and every line that can possibly be drawn in space satisfies the axioms. So how then can we know if the axioms always hold true? It is not enough that the axioms may *look* so simple and intuitive that they can be declared to be "obviously" true. It has been repeatedly shown that even the most seemingly simple and straightforward mathematical concepts, such as that of a "set," can produce contradictions that destroy the idea that a system is consistent. So we have to be wary of using simplicity and transparency as our sole guide in deciding on the truth of axioms.

One might wonder why we are so dependent on such a pedestrian method as applying each axiom to every element of the system in order to establish the truth of axioms and consequently the consistency of systems. Surely we can apply more powerful methods of reasoning to show whether a set of axioms is true even if it involves an infinite number of elements? One would think so except that Gödel proved that this could not be done except for very simple systems that do not cover most of the areas of interest to mathematicians. As soon as one goes beyond such trivial systems, one runs into problems because Gödel "proved that it is impossible to establish the internal logical consistency of a very large class of deductive systems—number theory, for example—*unless one adopts principles of reasoning so complex that their internal consistency is as open to doubt as that of the systems themselves*" (Nagel and Newman 2001, 5, my emphasis).

In other words, the price we pay for using more powerful reasoning methods to prove the consistency of axiomatic systems is that we lose transparency and simplicity in the rules of logic used in constructing those very methods, and now those rules cannot be assumed or shown to be valid. As the old saying goes, what you gain on the swings, you lose on the roundabouts. As a result, we arrive at Gödel's melancholy conclusion that can be stated as "no absolutely impeccable guarantee can be given that many significant branches of mathematical thought are entirely free of internal contradiction" (Nagel and Newman, 2001, 5). In other words, Gödel proved that the goal of proving consistency cannot be achieved *even in principle*.

This is quite a blow to the idea of determining absolute truth, because if we cannot show that a system is consistent, how can we depend upon its results, since lack of consistency means that a statement and its negation may both turn out to be provable within the system, and thus neither can be assumed to be true?

Our inability to show that an axiomatic system is consistent (i.e., free of contradictions as would be evidenced by the existence of two theorems each of which contradicts the other) is not the only problem. Gödel also showed that such systems are also *necessarily incomplete*. In other words, for all systems of interest, *there will always be some truths of that system that cannot be proven as theorems using only the axioms and rules of that system.* So the tantalizing goal that one day we might be able to develop a consistent system in which *every* true statement can be proven to be true also turns out to be a mirage. Neither completeness nor consistency is attainable.

The idea that there may be true statements that are unprovable has implications within mathematics because there are many general statements for which no exceptions have ever been found but their proofs have turned out to be elusive. Could they constitute some of the true statements that are unprovable? For example, Goldbach's Conjecture, which has been around since 1742, states that every even integer can be expressed as the sum of two prime numbers (where a prime number is defined as any number that is equal to or greater than 2 that is divisible only by itself or the number one, such as 2,3,5,7,11,13 . . .). No exception to this rule has ever been found. We can see that it holds true for small even integers ($8=3+5$, $10=3+7$, $12=5+7$, $24=11+13=7+17$, and so on) and with the advent of computers, this conjecture has been numerically checked and found to hold true for every even integer up to the staggeringly large number 4×10^{18}. But no general proof has yet been found. It is tempting to throw up one's hands and take refuge in the possibility that it may be one of those true but unprovable statements in number theory. But merely because it has been found to hold true up to such a large number is no guarantee that it will always hold true, another example of the well-known problem of induction. The very next even number might turn out to be the exception.

The danger of assuming that Goldbach's Conjecture is unprovable can be seen with Fermat's Last Theorem that states that no three positive integers a, b, and c can be found that satisfy the equation $a^n + b^n = c^n$ for any integer value of n greater than two. (For n=2, we have the famous Pythagorean theorem where such sets of integers do exist.) Although called a theorem, this was simply stated by Pierre de Fermat in 1637 and despite an annotation by him that he had found a proof of it, no proof was ever found amongst his papers. Attempts at proofs resulted in repeated failure so that it too was a favored candidate for being a true yet unprovable statement, even more so than Goldbach's Conjecture by virtue of its greater longevity. And yet, a proof did finally emerge in 1995, 358 years later.

In the efforts of mathematicians to make sure that their proofs lie solely within the abstract axiomatic system and not allow the use of everyday language to subtly introduce extraneous ideas that might lead them astray and lead to paradoxes, there have been attempts to drain the system of any everyday meanings. So, for example, the words "points" and "lines" used in geometry are simply *names* for elements that have specific properties within an axiomatic system and should not be conflated with how the words are used in ordinary language to denote entities in physical space. Bertrand Russell and Alfred North Whitehead in their monumental *Principia Mathematica* published in three volumes in 1910, 1912, and 1913 made an attempt to carry the abstraction of this program even further. The resulting system's elements are made up purely of symbols, the axioms consist of strings of those symbols, and the applicable logic consists of precise rules for manipulating those symbols so as to produce new strings of symbols. These new strings represent the theorems of the system, *with no meaning attached to them whatsoever.* Incidentally, because of this commitment to extreme abstract rigor, it took several hundred pages of such manipulation of symbols to arrive at the conclusion that is the equivalent of "1+1=2," where each term in the equation is just a symbol or a string of symbols.

Mathematical proofs have thus become disconnected from absolute truth claims. This is not to say that mathematicians do not care about creating systems that relate to the real world. They do, but some mathematicians care more about it than others, while some not at all. But many areas of highly abstract mathematics that started out having seemingly no applications in the real world have turned out to be extraordinarily useful for scientists, non-Euclidean geometry being a famous example. But usefulness is of secondary concern to many mathematicians. As Russell said, in the world of pure mathematics, the question of whether the axioms and the theorems derived from them are true is largely irrelevant. (What he calls "applied mathematics" is what we now refer to as theoretical physics or mathematical techniques like calculus.)

> Pure mathematics consists entirely of assertions to the effect that, if such and such a proposition is true of *anything*, then such and such another proposition is true of that thing. It is essential not to discuss whether the first proposition is really true, and not to mention what the anything is, of which it is supposed to be true. Both these points would belong to applied mathematics. We start, in pure mathematics, from certain rules of inference, by which we can infer that *if* one proposition is true, then so is some other proposition. These rules of inference constitute the major part of the principles of formal logic. We then take any hypothesis that seems amusing, and deduce its consequences. *If* our hypothesis is about *anything*, and not about some one or more particular things, then our deductions constitute mathematics. (Russell 1929, 75, emphasis in original)

This kind of purely symbolic structure disconnected from any real world applications has, as Russell quipped, resulted in pure mathematics becoming defined as "the subject in which we never know what we are talking about, nor whether what we are saying is true" (Russell 1929, 75). However Gödel's proof published in 1931 showed that Russell and Whitehead's ambitious goal of constructing a purely axiomatic system that would be both consistent and complete was unattainable.

D. Proof by logical contradiction that the square root of two is not a rational number (chapter 15)

This method of proof starts by assuming that the *negation* of a proposition is true and that this assumption leads to a logical contradiction, which implies that the proposition itself must be true. In this particular case, to prove that the square root of two is *not* a rational number, we start by assuming that it is and see where that leads.

A rational number is one that can be written as the ratio of two integers. For example, the number 1.5 is rational because it can be written as 6/4, 12/8, 3/2, and so on. Similarly, any number that terminates in a *finite* number of digits after the decimal point, such as 146.98, is a rational number because it can be written as 14698/100. A number with an infinite but *repeating* pattern of digits after the decimal point is also a rational number. For example, take the number 4.3151515 . . . where the sequence 15 is repeated indefinitely. Call this number x. If we multiply x by 10, we get $10x$=43.151515 . . . If we multiply x by 1000, we get $1000x$=4315.151515 . . . Subtracting $10x$ from $1000x$, we get $990x$=4272 *exactly*, since the repeating numbers after the decimal points are equal in both cases. Hence x=4272/990 exactly, and is thus a rational number. Similar reasoning can be applied to any number that has an infinitely repeating sequence. Conversely, the famous number π=3.1415927 . . . is *not* a rational number since it neither terminates nor does it contain a repeating pattern.

So if the square root of 2 is a rational number, then it can be written as the ratio a/b, where a and b are integers. We then make sure that the ratio has been "simplified" as much as possible by getting rid of all common factors in the numerator and denominator. For example, in the case of 146.98 discussed earlier, the ratio 14698/100 can be simplified to 7349/50 by cancelling out the only common factor that the numerator and denominator share, which is the number 2. In the case of 1.5, the ratio we would use is 3/2, since all the other ratios have common factors that when eliminated end up as 3/2 also.

So our starting assumption becomes that the square root of 2 is equal to a/b, where a and b are integers that do not have any common factors. We can now multiply each side by itself to get $2=a^2/b^2$. Hence $a^2=2b^2$. This implies that a^2 is an even number because it contains a factor of 2. But if the square of a number is even, that means the number itself must be even. Hence $a=2c$, where c is also an integer. This leads to $(2c)^2=2b^2$ and thus $b^2=2c^2$. This implies that b^2 is an even number and hence that b is also an even number. Thus b also has a factor 2. We have thus arrived at the conclusion that both a and b have the number 2 as a common factor, and this contradicts what we did at the start of the proof, where we got rid of all their common factors. We have thus arrived at a *logical* contradiction. Hence our starting assumption that the square root of 2 is a rational number must be false. Since there are only two possible alternatives (the square root of 2 is either rational or not rational), we can conclude that it is not rational.

E. Proof by logical contradiction that there is no largest prime number (chapter 15)

A prime number is any number that is 2 or larger that can only be divided by itself or the number 1. Starting from the smallest, they are 2, 3, 5, 7, 11, 13, 17, 19, 23, . . . Any number that is not a prime can be written as a product of prime numbers *in only one way* (this important property is called "the fundamental theorem of arithmetic"). So, for example, 90=2×3×3×5 and there is no other combination of primes that will make up 90. Any nonprime number can be divided by each of its constituent prime numbers.

The question is whether the series of prime numbers ends at some point or increases indefinitely. To prove that it doesn't end, we start with the opposite assumption that it does, and that this largest prime number is P. We then construct a new number Q consisting of the product of all the prime numbers plus 1. That is,

$$Q = 2 \times 3 \times 5 \times 7 \times 11 \times 13 \times 17 \times 19 \times 23 \times .... \times P + 1$$

The number Q is clearly larger than P. Hence according to our starting assumption, Q cannot be a prime number and thus must be a product of prime numbers. But we see that Q is not divisible by *any* of the prime numbers up to P because if we divide Q by each of them, we always get a remainder of 1. Hence either Q is itself a prime number larger than P or it must be divisible by a prime number that is larger than P. Either way, this contradicts our starting assumption that a largest prime number P exists. Hence there can be no largest prime number.

F. Conservation laws in elementary particle physics (chapter 16)

In addition to energy conservation, two other strict conservation laws that govern our daily lives are those for momentum and angular momentum. But conservation laws play immensely important roles in many areas of science that are far removed from direct sensory experience. In elementary particle physics, for example, one encounters many conservation laws, such as baryon number and lepton number. Baryons and leptons are two classes of particles and the total number of particles in each class remains the same in any reaction. Protons and neutrons are examples of baryons and electrons, muons, and neutrinos are examples of leptons.

There is also a conservation law involving CPT, where C stands for "charge conjugation," P stands for "parity," and T for "time reversal." Conservation under charge conjugation means that if we replace every particle involved in a reaction by its corresponding anti-particle, the reaction would proceed identically. Conservation of parity means that we would not be able to tell if an elementary particle reaction we observe was seen by us directly or was observed reflected in a mirror, where left and right-handedness would be switched. Conservation under time reversal means that we could not tell if the film of a reaction that was being shown to us was running forward or backward. What CPT conservation tells us is that in any reaction, we would get the same result when all three operations C, P, and T are applied to the particles involved in the reaction. CPT conservation is thought to be strictly true under all circumstances. This is not the case for C, P, and T separately or in any other combination. At various times, each one was thought to be conserved too but over time violations of those conservation laws were found.

All conservations laws are universal claims. As an example of how the contradiction of a universal claim is established by means of establishing an existence claim, it used to be thought that in addition to CPT conservation, just CP alone (i.e., a combination of just charge conjugation and parity) was also conserved in every reaction involving elementary particles. Why did we believe this universal claim? Because no reaction violating it had ever been seen. Postulating the violation of CP-conservation required finding a reaction that violated it, thus constituting a new existence claim. Some scientists suspected that it might be violated under certain conditions and one such rare reaction was detected in 1964, which was confirmed in subsequent experiments. It was only then that the universal claim that CP was always conserved was accepted as not being always true. Now the new universal claim is that CP is conserved *except for this particular kind of reaction that will violate CP every time it occurs.*

There have also long been suggestions based on some particle theories that baryon number conservation can be violated under certain conditions but such a violation has not been seen yet. That existence claim is still being investigated but until a clear-cut case is found, the universal baryon number conservation law will be assumed to hold.

G. How the enemies of science misuse science epistemology (chapter 19)

The arguments and methods used by groups that oppose any regulatory action to fight global warming and climate change by limiting and even reducing the emission of greenhouse gases are the same (and sometimes involve the very same people and organizations) that tried in the past to discredit the consensus views of scientists that smoking and second-hand smoke cause cancer, that acid rain is bad for the environment, that chlorofluorocarbons (CFCs) were creating holes in the ozone layer, and the dangers of strategic missile systems. Their goal is to exploit the slivers of doubt that always exist in science, and which honest scientists always acknowledge, to cast doubt on the expert consensus views of scientists

The methods they adopt are very similar in every case. One is to do their own "research" in their own laboratories or fund scientists to do research, and then selectively use any result that they think is favorable to their case, however trivial, to loudly trumpet that they have contradicted the consensus views of experts.

They will do this even when their own internal studies go counter to their public statements. Geoffrey Supran and Naomi Oreskes reveal, for example, that there was a wide divergence between what the internal documents of the oil giant ExxonMobil said on the issue of climate change and what their public stance was, with the latter seeking to exaggerate the level of doubt.

> We find that as documents become more publicly accessible, they increasingly communicate doubt. This discrepancy is most pronounced between advertorials and all other documents. For example, accounting for expressions of reasonable doubt, 83% of peer-reviewed papers and 80% of internal documents acknowledge that climate change is real and human-caused, yet only 12% of advertorials do so, with 81% instead expressing doubt. We conclude that ExxonMobil contributed to advancing climate science—by way of its scientists' academic publications—but promoted doubt about it in advertorials. Given this discrepancy, we conclude that ExxonMobil misled the public. (Supran and Oreskes 2017)

The first effort of those seeking to undermine science's credibility using public relations techniques to cast doubt was in the 1950s on behalf of the cigarette industry, even though the dangers of cigarette smoking had been well known decades earlier.

> German scientists had shown in the 1930s that cigarette smoking caused lung cancer, and the Nazi government had run major antismoking campaigns; Adolf Hitler forbade smoking in his presence. However, the German scientific work was tainted by its Nazi associations, and to some extent ignored, if not actually suppressed, after the war; it had taken some time to be rediscovered and independently confirmed. Now, however, American researchers—not Nazis—were calling the matter "urgent," and the news media were reporting it. "Cancer by the carton" was not a slogan the tobacco industry would embrace.
> The tobacco industry was thrown into panic. One industry memo noted that their salesmen were "frantically alarmed." So industry executives made a fateful decision, one that would later become the basis on which a federal judge would find the industry guilty of conspiracy to commit fraud—a massive and ongoing fraud to deceive the American public about the health effects of smoking. The decision was to hire a public relations firm to challenge the scientific evidence that smoking could kill you. (Oreskes and Conway 2010, 15)
> . . .
> The industry made its case in part by cherry-picking data and focusing on unexplained or anomalous details. No one in 1954 would have claimed that everything that needed to be known about smoking and cancer was known,

and the industry exploited this normal scientific honesty to spin unreasonable doubt. (Oreskes and Conway 2010, 18)

The template of seeking to use the lack of certainty and proof in science to falsely create the impression of extreme uncertainty and doubt has to be countered by making the public aware that one can always construct alternative theories and there will always be anomalous data, but that does not mean that a sound reasoned judgment cannot be arrived at nor that all theories are equally credible. Scientific consensus comes down on the side of the argument where the *preponderance of evidence* lies, and this is why that consensus should be preferred as the basis for action.

Acknowledgments

A book such as this is never the product of a single person. It is only possible because of the assistance provided by many others and it is a pleasure for me to acknowledge and thank them here, though that cannot in any way repay my debt to them.

Many thanks are due to R. John Leigh, who from the very inception of this project expressed interest and enthusiasm for the ideas I was proposing, read chapters as they were written, and gave valuable criticisms. His healthy skepticism, clearly expressed, enabled me to address many weaknesses and lacunae in the early stages.

It is a pleasure to acknowledge Barbara Burgess-Van Aken from the English department of Case Western Reserve University, who read an early draft of the full manuscript and provided so many helpful suggestions. In addition to coteaching a seminar course with me on the history and philosophy of science, she and another CWRU colleague, Arthur Evenchik, taught me a lot about what makes for good writing and how to evaluate it.

The following people read all or part of early versions of the manuscript and provided helpful feedback and encouragement: Danny Solow, Harris Taylor, Peter Whitehouse, Suren Fernando, Nimal Gunatilleke, and Indrakumar Jayawardene. Many erudite and knowledgeable readers and commenters on my blog at https://freethoughtblogs.com/singham/ contributed many pieces of information, sharp insights, and informed critiques of various ideas in this book that I tentatively broached there. Craig F. Bohren proved to be a valuable source of information on the literature, pointing me to books and articles on the history of science. Jeff Hess and Brad Ricca helped me prepare my book manuscript proposal.

The anonymous reviewers selected by Oxford University Press to evaluate my manuscript provided me with many helpful comments and suggestions for improvement.

It is a pleasure to thank Josh White, who took my crude sketches of the four figures in the book and made them much clearer and pleasing to look at, with the three tree figures obtained using the free online open-source software called Malsys, available at http://www.malsys.cz and that uses an L-system generator, which is a common technique for generating fractal structures.

And I must acknowledge the deep debt of gratitude to Vice Provost Donald Feke for his constant support during my time teaching at the university and after

I retired, and to the excellent library and librarians at CWRU who provided me with all the resources I needed.

Finally, it is a pleasure to thank my editor at Oxford University Press, Jeremy Lewis. His support and encouragement during the entire process and his prompt responses to my many questions made the process of getting the manuscript into print much more pleasurable.

References

Abbott, B. P., R. Abbott, T. D. Abbott, M. R. Abernathy, F. Acernese, K. Ackley, C. Adams, T. Adams, P. Addesso, R. X. Adhikari, et al. 2016. "Observation of Gravitational Waves from a Binary Black Hole Merger." *Physical Review Letters* 116, 061102–061118 (February). https://doi.org/10.1103/PhysRevLett.116.061102.

Ade, P. A. R., N. Aghanim, M. Arnaud, M. Ashdow, J. Aumont, C. Baccigalupi, A. J. Banday, et al. 2016. "Planck 2015 Results: XIII. Cosmological Parameters." *Astronomy & Astrophysics* 594, A13 (October). https://doi.org/10.1051/0004-6361/201525830.

Afshordi, Niayesh, and João Magueijo. 2016. "Critical Geometry of a Thermal Big Bang." *Physical Review D* 94, no. 10 (November 15): 101301-101301-6. https://doi.org/10.1103/PhysRevD.94.101301.

Akerib, D. S., S. Alsum, H. M. Araújo, X. Bai, A. J. Bailey, J. Balajthy, P. Beltrame, et al. 2017. "Results from a Search for Dark Matter in the Complete LUX Exposure." *Physical Review Letters* 118, no. 2 (January 13): 021303-1-021303-8. https://doi.org/10.1103/PhysRevLett.118.021303.

Alecci, Scilla. 2018. "Undercover Reporters Expose 'Bogus' Scientific Conferences." *International Consortium of Investigative Journalists*, September 12, 2018. https://www.icij.org/blog/2018/09/undercover-reporters-expose-bogus-scientific-conferences/.

Arehart, Mark. 2016. "Gas, Electric Or Steam? Car Shopping, 100 Years Ago." *NPR's All Things Considered*, September 12. http://www.npr.org/2016/09/12/492841796/gas-electric-or-steam-car-shopping-100-years-ago.

Bacon, Francis. 1939 [1620]. *Novum Organum*, section XLVI. In *The English Philosophers from Bacon to Mill*, edited by E. A. Burtt. New York: Random House. http://www.constitution.org/bacon/nov_org.htm.

Ball, Philip. 2012. *Curiosity: How Science Became Interested in Everything.* Chicago: University of Chicago Press.

Ball, Philip. 2018. *Beyond Weird: Why Everything You Thought You Knew About Quantum Physics is Different.* Chicago: University of Chicago Press.

Ballentine, L. E. 1970. "The Statistical Interpretation of Quantum Mechanics." *Reviews of Modern Physics* 42, no. 4 (October): 358–81. https://doi.org/10.1103/RevModPhys.42.358.

Barnes, Barry. 1982. *T. S. Kuhn and Social Science.* New York: Columbia University Press.

Behe, Michael. 1996. *Darwin's Black Box.* New York: Free Press.

Bell, J. S. 1964. "On the Einstein-Podolsky-Rosen Paradox." *Physics* 1, 195–200. Reprinted in J. S. Bell, *Speakable and Unspeakable in Quantum Mechanics*, 14–21. New York: Cambridge University Press, 1987.

Bell, J. S. 1976. "How to Teach Special Relativity." *Progress in Scientific Culture* 1, no. 2 (Summer). Reprinted in Bell, *Speakable and Unspeakable in Quantum Mechanics*, 67–80.

Bell, J. S. 1981. "Quantum Mechanics for Cosmologists." In *Quantum Gravity 2*, edited by C. Isham, R. Penrose, and D. Sciama, 611–37. Oxford: Clarendon Press. Reprinted in Bell, *Speakable and Unspeakable in Quantum Mechanics*, 117–38.

Bell, J. S. 1986. "Six Possible Worlds of Quantum Mechanics." *Proceedings of the Nobel Symposium 65: Possible Worlds in Arts and Science,* edited by Sture Allen, Stockholm (August 11–15). Reprinted in Bell, *Speakable and Unspeakable in Quantum Mechanics,* 181–95.

Bell, J. S. 1987. "Are There Quantum Jumps?" In *Schrodinger, Centenary Celebration of a Polymath,* edited by C. W. Kilmister, 41–52. New York: Cambridge University Press. Reprinted in Bell, *Speakable and Unspeakable in Quantum Mechanics,* 201–12.

Beller, Mara. 1999. *Quantum Dialogue: The Making of a Revolution.* Chicago: University of Chicago Press.

Bethe, H. A. 1939. "Energy Production in Stars." *Physical Review* 55, no. 5 (March 1): 434–56. https://doi.org/10.1103/PhysRev.55.434.

Bethe, H. A., and C. L. Critchfield. 1938. "On the Formation of Deuterons by Proton Combination." *Physical Review* 54, no. 10 (November 15): 862. https://doi.org/10.1103/PhysRev.54.862.2.

Blackmore, Susan. 2005. *Consciousness: A Very Short Introduction.* Oxford: Oxford University Press.

Bloor, David. 1991. *Knowledge and Social Imagery* (second edition). Chicago: University of Chicago Press.

Boehm, Christopher. 2012. *Moral Origins: The Evolution of Virtue, Altruism, and Shame.* New York: Basic Books.

Bohm, David. 1957. *Causality and Chance in Modern Physics.* Philadelphia: University of Pennsylvania Press.

Brasier, A. T., D. McIlroy, and N. McLoughlin (eds.). 2014. "Remarkable Preservation of Brain Tissues in an Early Cretaceous Iguanodontian Dinosaur." *Earth System Evolution and Early Life: A Celebration of the Work of Martin Brasier.* Geological Society, London, Special Publications 448, 383–98. http://doi.org/10.1144/SP448.3.

Brown, Kevin. 2017. *Reflections on Relativity.* http://www.mathpages.com/rr/rrtoc.htm.

Burchfield, Joe D. 1975. *Lord Kelvin and the Age of the Earth.* New York: Science History Publication.

Burdick, Alan. 2017. " 'Paging Dr. Fraud': The Fake Publishers That Are Ruining Science." *The New Yorker,* March 22. https://www.newyorker.com/tech/annals-of-technology/paging-dr-fraud-the-fake-publishers-that-are-ruining-science.

Butterfield, Herbert. 1931. *The Whig Interpretation of History.* London: Bell.

Butterfield, Herbert. 1957. *The Origins of Modern Science 1300–1800.* New York: MacMillan.

Byrne, Peter. 2007. "The Many Worlds of Hugh Everett." *Scientific American* 297, no. 6, December. https://www.scientificamerican.com/magazine/sa/2007/12-01/.

Calaprice, Alice. 2005. *The New Quotable Einstein.* Princeton, NJ: Princeton University Press.

Catling, David C. 2013. *Astrobiology: A Very Short Introduction.* Oxford: Oxford University Press.

Chalmers, Alan. 1990. *Science and Its Fabrication.* Minneapolis: University of Minnesota Press.

Chang, Kenneth. 2004. "Researcher Loses Ph.D. Over Discredited Papers." *New York Times,* June 15. http://www.nytimes.com/2004/06/15/science/researcher-loses-phd-over-discredited-papers.html.

Chiang, Ted. 2016. *Stories of Your Life and Others.* New York: Vintage Books.

Cho, Adrian. 2012. "Once Again, Physicists Debunk Faster-Than-Light Neutrinos." *Science*, June 8. http://www.sciencemag.org/news/2012/06/once-again-physicists-debunk-faster-light-neutrinos.

Chomsky, Noam. 1975. *Reflections on Language*. New York: Pantheon.

Chown, Marcus. 2018. "Stellar Prophesies: The Power of Prediction." *New Humanist*, Autumn. https://newhumanist.org.uk/articles/5362/stellar-prophesies-the-power-of-prediction.

Cockell, Charles S. 2017. "The Laws of Life." *Physics Today* 70, no. 3 (March): 42–48. https://doi.org/10.1063/PT.3.3493.

Collins, H. M. 1983. "An Empirical Relativist Programme in the Sociology of Scientific Knowledge." In *Science Observed: Perspective on the Social Study of Science*, edited by Karin D. Knorr-Cetina and Michael Mulkay, 85–113. London: SAGE Publications.

Cook, Alan. 2000. "Success and Failure in Newton's Lunar Theory." *Astronomy & Geophysics* 41, no. 6 (December 1): 6.21–6.25. https://doi.org/10.1046/j.1468-4004.2000.41621.x.

Cramer, John G. 1986. "The Transactional Interpretation of Quantum Mechanics." *Reviews of Modern Physics* 58, no. 3, (July): 647–87. https://doi.org/10.1103/RevModPhys.58.647.

Cutler, Alan. 2003. *The Seashell on the Mountaintop: A Story of Science Sainthood and the Humble Genius Who Discovered a New History of the Earth*. New York: Dutton.

Cyranoski, David. 2006. "Verdict: Hwang's Human Stem Cells Were All Fakes." *Nature* 439, no. 7073 (January 12): 122–23. https://doi.org/10.1038/439122a.

Dalrymple, G. Brent. 1991. *The Age of the Earth*. Stanford, CA: Stanford University Press.

Danielson, Dennis R. 2001. "The Great Copernican Cliché." *American Journal of Physics* 69 (October): 1029–35. https://doi.org/10.1119/1.1379734.

Darwin, Charles. 1859. *On the Origin of Species*. London: John Murray.

Darwin, Charles. 1872. *The Origin of Species* (sixth edition). New York: The New American Library.

Dennett, Daniel C. 1991. *Consciousness Explained*. New York: Little, Brown and Company.

Dennett, Daniel C. 2017. *From Bacteria to Bach and Back: The Evolution of Minds*. New York: Norton.

Desmond, Adrian, and James Moore. 1991. *Darwin: The Life of a Tormented Evolutionist*. New York: Norton.

Duhem, Pierre. 1954 [1906]. *The Aim and Structure of Physical Theory*. Translated by Philip P. Wiener. Princeton, NJ: Princeton University Press.

Einstein, Albert. 1949. "Remarks to the Essays Appearing in this Collective Volume." In *Albert Einstein: Philosopher-Scientist*, 663–88. Evanston, IL: Library of Living Philosophers. https://www.marxists.org/reference/archive/einstein/works/1940s/reply.htm.

England, Philip, Peter Molnar, and Frank Richter. 2007. "John Perry's Neglected Critique of Kelvin's Age for the Earth: A Missed Opportunity in Geodynamics." *GSA Today* 17, no. 1, 4–9. https://doi.org/10.1130/GSAT01701A.1.

Farrington, Benjamin. 1964. *The Philosophy of Francis Bacon*. Liverpool: Liverpool University Press.

Feyerabend, Paul. 1993. *Against Method: Outline of an Anarchist Theory of Knowledge*. New York: Verso.

Feynman, Richard P. 1985. *QED: The Strange Theory of Light and Matter*. Princeton, NJ: Princeton University Press.

Feynman, Richard P. 1998. *The Meaning of It All: Thoughts of a Citizen Scientist.* Reading: Perseus Books.

French, Christopher C., and Anna Stone. 2014. *Anomalistic Psychology: Exploring Paranormal Belief and Experience.* Basingstoke: Palgrave Macmillan.

Frova, Andrea, and Mariapiera Marenzana. 2006. *Thus Spoke Galileo: The Great Scientist's Ideas and Their Relevance to the Present Day.* Oxford: Oxford University Press.

Gelman, Andrew. 2011. "Induction and Deduction in Bayesian Data Analysis." RMM 2, 67–78, *Special Topic: Statistical Science and Philosophy of Science*, edited by Deborah G. Mayo, Aris Spanos, and Kent W. Staley. http://www.stat.columbia.edu/~gelman/research/published/philosophy_online4.pdf.

Gelman, Andrew, and Cosma Rohilla Shalizi. 2013. "Philosophy and the Practice of Bayesian Statistics." *British Journal of Mathematical and Statistical Psychology* 66, no. 1 (February 1): 8–38. https://doi.org/10.1111/j.2044-8317.2011.02037.x.

Giere, Ronald N., John Bickle, and Robert Mauldin. 2006. *Understanding Scientific Reasoning* (5th edition). Belmont, CA: Wadsworth.

Grant, Edward. 2004. *Science and Religion 400 B.C. to A. D. 1550: From Aristotle to Copernicus.* Baltimore: Johns Hopkins University Press.

Gross, Charles. 2011. "Disgrace: On Marc Hauser." *The Nation*, December 21. https://www.thenation.com/article/disgrace-marc-hauser/.

Gross, Paul R., and Norman Levitt. 1994. *Higher Superstition: The Academic Left and Its Quarrels with Science.* Baltimore: Johns Hopkins University Press.

Gross, Paul R., Norman Levitt, and Martin W. Lewis. 1996. *The Flight From Science and Reason*, Annals of the New York Academy of Sciences, 775. New York: New York Academy of Sciences.

Haldane, J. B. S. 1934. *Fact and Faith.* London: Watts and Company.

Haltiwanger, John. 2017. "Fear of Vampires in Malawi Leads to Five Deaths, Forces Out U.N." *Newsweek*, October 9. http://www.newsweek.com/fear-vampires-malawi-kills-five-forces-un-out-681030.

Harding, Sandra. 2006. *Science and Social Inequality: Feminist and Postcolonial Issues.* Urbana: University of Illinois Press.

Hawking, Stephen, and Leonard Mlodinow. 2010. *The Grand Design.* New York: Bantam Books.

Haynes, Korey. 2018. "The Jesuit Astronomer Who Conceived of the Big Bang." *Astronomy*, October 12. http://astronomy.com/news/2018/10/the-jesuit-astronomer-who-conceived-of-the-big-bang.

Hazen, Robert M. 2005. *Genesis: The Scientific Quest for Life's Origin.* Washington, DC: Joseph Henry Press.

Hedges, S. Blair, and Sudhir Kumar (eds.). 2009. *The Timetree of Life.* Oxford: Oxford University Press.

Henig, Robin Marantz. 2000. *The Monk in the Garden.* Boston: Houghton Mifflin.

Hill, Austin Bradford. 1965. "The Environment and Disease: Association or Causation." *Proceedings of the Royal Society of Medicine* 58, no. 5 (May): 295–300. https://www.ncbi.nlm.nih.gov/pmc/articles/PMC1898525/.

Hoffman, Andrew J. 2018. "Why the Web Has Challenged Scientists' Authority—and Why They Need to Adapt." *The Conversation*, accessed October 14. https://theconversation.com/why-the-web-has-challenged-scientists-authority-and-why-they-need-to-adapt-91893.

Hofstadter, Douglas R. 1999. *Gödel, Escher, Bach: an Eternal Golden Braid.* New York: Basic Books.

Hoke, Frankin. 1995. "Scientists See Broad Attack Against Research And Reason." *The Scientist* 9, no.14 (July 10): 1. https://www.the-scientist.com/?articles.view/articleNo/ 17474/title/Scientists-See-Broad-Attack-Against-Research-And-Reason/.

Horgan, John. 1996. *The End of Science*. Reading: Addison-Wesley.

Horgan, John. 2014. "Physics Titan Still Thinks String Theory Is 'On the Right Track.'" *Scientific American*, September 22. https://blogs.scientificamerican.com/cross-check/ physics-titan-still-thinks-string-theory-is-on-the-right-track/.

Howard, Don A. 2017. "Einstein's Philosophy of Science." In *The Stanford Encyclopedia of Philosophy* (Fall), edited by Edward N. Zalta. https://plato.stanford.edu/archives/ fall2017/entries/einstein-philscience/.

Howson, Colin, and Peter Urbach. 1993. *Scientific Reasoning: The Bayesian Approach*. Chicago: Open Court.

Hume, David. 1779. *Dialogues Concerning Natural Religion*, quoted in "Hume on Religion" by Paul Russell and Anders Kraal, in *The Stanford Encyclopedia of Philosophy* (Summer 2017 Edition), edited by Edward N. Zalta. https://plato.stanford.edu/entries/ hume-religion/#WasHumAth.

Iliffe, Rob. 2017. *Priest of Nature: The Religious Worlds of Isaac Newton*. New York: Oxford University Press.

Jackson, Patrick Wyse. 2006. *The Chronologers' Quest: The Search for the Age of the Earth*. New York: Cambridge University Press.

Jaki, Stanley L. 1978. "Johann Georg von Soldner and the Gravitational Bending of Light, with an English Translation of His Essay on It Published in 1801." *Foundations of Physics* 8, nos. 11/12: 927–50. https://doi.org/10.1007/BF00715064.

Jammer, Max. 1966. *The Conceptual Development of Quantum Mechanics*. New York: McGraw-Hill.

Jenkin, Fleeming. 1867. "Review of The Origin of Species." *North British Review* 46 (June): 277—318.

Jenkins, Stephen H. 2004. *How Science Works: Evaluating Evidence in Biology and Medicine*. New York: Oxford University Press.

Johnson, Phillip. 1991. *Darwin on Trial*. Washington, DC: Regnery.

Katz, Leonard D. (ed.). 2000. *Evolutionary Origins of Morality: Cross Disciplinary Perspectives*. Bowling Green, OH: Imprint Academic.

Keller, Evelyn Fox, and Helen E. Longino (eds.). 1996. *Feminism and Science*. Oxford: Oxford University Press.

Kitcher, Philip. 1993. *The Advancement of Science: Science without Legend, Objectivity without Illusions*. New York: Oxford University Press.

Kolb, David A. 1984. *Experiential Learning*. Englewood Cliffs, NJ: Prentice-Hall.

Kolb, Edward W., Sabino Matarrese, and Antonio Riotto. 2006. "On Cosmic Acceleration without Dark Energy." *New Journal of Physics* 8, no. 12: 322. https://doi.org/10.1088/ 1367-2630/8/12/322.

Kosso, Peter. 1992. *Reading the Book of Nature: An Introduction to the Philosophy of Science*. Cambridge: Cambridge University Press.

Krebs, Dennis. 2011. *The Origins of Morality: An Evolutionary Account*. New York: Oxford University Press.

Kroupa, Pavel. 2016. "Has Dogma Derailed the Scientific Search for Dark Matter?" *Aeon*, November 25. https://aeon.co/ideas/has-dogma-derailed-the-scientific-search-for-dark-matter.

Kuhn, Thomas S. 1957. *The Copernican Revolution: Planetary Astronomy in the Development of Western Thought*. Cambridge, MA: Harvard University Press.

Kuhn, Thomas S. 1970. *The Structure of Scientific Revolutions*. Chicago: University of Chicago Press.

Ladyman, James. 2002. *Understanding Philosophy of Science*. London: Routledge.

Lakatos, Imre. 1970. "Falsification and the Methodology of Scientific Research Programmes." In *Criticism and the Growth of Knowledge*, edited by I. Lakatos and A. Musgrave, 91–196. Cambridge: Cambridge University Press. Reprinted in *The Methodology of Scientific Research Programmes*, edited by John Worrall and Gregory Currie, 8–101, Cambridge: Cambridge University Press, 1986.

Lakatos, Imre. 1973. "Science and Pseudoscience." In *The Methodology of Scientific Research Programmes*, edited by John Worrall and Gregory Currie, 1–7, 1986. Cambridge: Cambridge University Press.

Lane, Nick. 2002. "Born to the Purple: the Story of Porphyria." *Scientific American*, December 16. https://www.scientificamerican.com/article/born-to-the-purple-the-st/.

Larson, Edward J. 1997. *The Summer of the Gods*. Cambridge, MA: Harvard University Press.

Larson, Edward J., and Larry Witham. 1998. "Leading Scientists Still Reject God." *Nature* 394, no. 6691 (July 23): 313. https://doi.org/10.1038/28478.

Laser Interferometer Gravitational-Wave Observatory (LIGO). 2016. "Gravitational Waves Detected 100 Years After Einstein's Prediction." Ligo News Release, February 11. https://www.ligo.caltech.edu/news/ligo20160211.

Latour, Bruno, and Steve Woolgar. 1979. *Laboratory Life: The Social Construction of Scientific Facts*. Beverly Hills, CA: SAGE Publications.

Laudan, Larry. 1977. *Progress and its Problems: Towards a Theory of Scientific Growth*. Berkeley: University of California Press.

Laudan, Larry. 1983. "The Demise of the Demarcation Problem." In *Physics, Philosophy, and Psychoanalysis*, edited by R. S. Cohen and L. Laudan, 111–27. Dordrecht: D. Reidel Publishing Company. Reprinted in *But Is It Science?*, edited by Michael Ruse, 337–50, 1986. Amherst: Prometheus Books.

Laudan, Larry. 1984. *Science and Values*. Berkeley: University of California Press.

Laudan, Larry 1990. *Science and Relativism: Some Key Controversies in the Philosophy of Science*. Chicago: University of Chicago Press.

Laughlin, R. B., and David Pines. 2000. "The Theory of Everything." *Proceedings of the National Academy of Sciences* 97, no.1 (January 4): 28–31. https://doi.org/10.1073/pnas.97.1.28

Lederman, Muriel, and Ingrid Bartsch (eds.). 2001. *The Gender and Science Reader*. London: Routledge.

Lemonick, Michael D. 2014. "Q&A: The 5 Ingredients Needed for Life Beyond Earth." *National Geographic*, June 26. http://news.nationalgeographic.com/news/2014/06/140625-kepler-exoplanets-life-astrobiology-goldilocks-nasa/.

Levenson, Thomas. 2015. *The Hunt for Vulcan: And How Albert Einstein Destroyed a Planet, Discovered Relativity, and Deciphered the Universe*. New York: Random House.

Lewontin, Richard C. 1983. "Introduction." In *Scientists Confront Creationism*, edited by Laurie R. Godfrey, xxiii–xxvi. New York: Norton.

Lisle, Jason. 2007. "Does Distant Starlight Prove the Universe Is Old?" *Answers in Genesis*, https://answersingenesis.org/astronomy/starlight/does-distant-starlight-prove-the-universe-is-old/.

Livio, Mario. 2013. *Brilliant Blunders*. New York: Simon & Schuster.

Lloyd, Seth. 2006. In *What We Believe but Cannot Prove: Today's Leading Thinkers on Science in the Age of Certainty*, edited by John Brockman, 55–56. New York: Harper Collins. https://www.amazon.com/What-Believe-but-Cannot-Prove/dp/0060841818.

Longino, Helen E. 1990. *Science as Social Knowledge: Values and Objectivity in Scientific Inquiry*. Princeton, NJ: Princeton University Press.

Marcum, James A. 2018. "The Revolutionary Ideas of Thomas Kuhn." *The Times Literary Supplement*, January 17. https://www.the-tls.co.uk/articles/public/scientific-revolutions-thomas-kuhn/.

Mayr, Ernst. 1990. "When is Historiography Whiggish?" *Journal of the History of Ideas* 51, no. 2 (April–June): 301–309. https://doi.org/10.2307/2709517.

McKay, Christopher P. 2014. "Requirements and Limits for Life in the Context of Exoplanets." *Proceedings of the National Academy of Sciences* 111, no. 35 (September 2): 12628–33. https://doi.org/10.1073/pnas.1304212111.

Medawar, Peter B. 1964. "Is the Scientific Paper Fraudulent? Yes; It Misrepresents Scientific Thought." *The Saturday Review* (August 1): 42–43. http://www.unz.org/Pub/SaturdayRev-1964aug01-00042.

Milgrom, Mordehai. 2002. "MOND—Theoretical Aspects." *New Astronomy Reviews* 46, Issue 12 (November): 741–53. https://doi.org/10.1016/S1387-6473(02)00243-9.

Milgrom, Mordehai. 2014. "The MOND Paradigm of Modified Dynamics." *Scholarpedia* 9, no. 6: 31410. https://doi.org/10.4249/scholarpedia.31410.

Miller, Kenneth R. 1999. *Finding Darwin's God*. New York: Harper Collins.

Morris, Simon Conway. 1998. *The Crucible of Creation*. Oxford: Oxford University Press:

Morris, Susan W. 1994. "Fleeming Jenkin and 'The Origin of Species': A Reassessment." *British Journal for the History of Science* 27, no. 3 (September): 313–43. https://www.jstor.org/stable/4027601.

Murzi, Mauro. n.d. "Rudolf Carnap (1891—1970)." *Internet Encyclopedia of Philosophy*. http://www.iep.utm.edu/carnap.

Nagel, Ernest, and James R. Newman. 2001. *Gödel's Proof*. New York: New York University Press.

Nevin, Rick. 2007. "Understanding International Crime Trends: The Legacy of Preschool Lead Exposure." *Environmental Research* 104, no. 3 (July): 315–36. https://doi.org/10.1016/j.envres.2007.02.008.

Numbers, Ronald L. 1992. *The Creationists: The Evolution of Scientific Creationism*. New York: Knopf.

Okasha, Samir. 2016. *Philosophy of Science: A Very Short Introduction*. Oxford: Oxford University Press.

Oppy, Graham. 2017. "Ontological Arguments." *The Stanford Encyclopedia of Philosophy* (Summer 2017 Edition), edited by Edward N. Zalta. https://plato.stanford.edu/archives/sum2017/entries/ontological-arguments/.

Oreskes, Naomi, and Erik M. Conway. 2010. *Merchants of Doubt*. New York: Bloomsbury Press.

Pace, Norman R., Jan Sapp, and Nigel Goldenfeld. 2012. "Phylogeny and Beyond: Scientific, Historical, and Conceptual Significance of the First Tree of Life." *Proceedings of the National Academy of Sciences* 109, no. 4 (January 24): 1011–18. https://doi.org/10.1073/pnas.1109716109.

Palci, Alessandro. 2016. "Why Are We Still Searching for the Loch Ness Monster?" *The Conversation*, April 25. https://theconversation.com/why-are-we-still-searching-for-the-loch-ness-monster-58157.

Paley, William. 1802. *Natural Theology*. Philadelphia: John Morgan.

Patterson, Roger 2015. "Were You There? Pointing to God as Creator." *Answers in Genesis*, https://answersingenesis.org/evidence-against-evolution/were-you-there/.

Paulos, John Allen. 2008. *Irreligion: A Mathematician Explains Why the Arguments for God Just Don't Add Up*. New York: Hill and Wang.

Pennock, Robert T. 1999. *Tower of Babel: The Evidence against the New Creationism*. Cambridge, MA: MIT Press.

Perry, J. 1895a. "On the Age of the Earth." *Nature* 51, no. 1314 (January 3): 224–27. https://doi.org/10.1038/051224a0.

Perry, J. 1895b. "On the Age of the Earth." *Nature* 51, no. 1319 (February 7): 341–42. https://doi.org/10.1038/051341b0.

Perry, J. 1895c. "On the Age of the Earth." *Nature* 51, no. 1329 (April 18): 582–85. https://doi.org/10.1038/051582a0.

Pew Research Center for the People & the Press. 2009. "Public Praises Science; Scientists Fault Public, Media." Pew Research Center, July 9. http://www.people-press.org/2009/07/09/public-praises-science-scientists-fault-public-media/.

Pigliucci, Massimo. 2010. *Nonsense on Stilts: How to Tell Science from Bunk*. Chicago: University of Chicago Press.

Popper, Karl R. 1965. *Conjectures and Refutations: The Growth of Scientific Knowledge*. New York: Harper Torchbooks.

Popper, Karl R. 1968a. *The Logic of Scientific Discovery*. London: Hutchinson.

Popper, Karl. R. 1968b. "Is There an Epistemological Problem of Perception?" In *Problems in the Philosophy of Science*, edited by Imre Lakatos and Alan Musgrave, 163. Amsterdam: North-Holland.

Principe, Lawrence M. 2011. "Alchemy Restored." *Isis* 102, no. 2 (June): 305–12. https://doi.org/10.1086/660139.

Pross, Addy. 2012. *What is Life?* Oxford: Oxford University Press.

Putnam, H. 1975. "What is Mathematical Truth?" In *Mathematics, Matter and Method: Philosophical Papers*, Volume 1, 60–78. Cambridge: Cambridge University Press.

Rajantie, Arttu. 2016. "The Search for Magnetic Monopoles." *Physics Today* 69, no. 10 (October): 40–46. https://doi.org/10.1063/PT.3.3328.

Rayleigh, Lord. 1921. "The Age of the Earth." *Nature* 108, no. 2713 (October 27): 279–81. https://doi.org/10.1038/108279a0.

Reyes, Jessica Wolpaw. 2007. "Environmental Policy as Social Policy? The Impact of Childhood Lead Exposure on Crime." Working Paper 13097, National Bureau of Economic Research, May. http://www.nber.org/papers/w13097.

Ricardo, Alonso, and Jack W. Szostak. 2009. "The Origin of Life on Earth: Fresh Clues Hint at How the First Living Organisms Arose from Inanimate Matter." *Scientific American* 301 (September): 54–61. https://www.scientificamerican.com/article/origin-of-life-on-earth/.

Rovelli, Carlo. 2016. *Reality is Not What it Seems: The Journey to Quantum Gravity*. New York: Riverhead Books.

Rubin, Vera C., and W. Kent Ford Jr. 1970. "Rotation of the Andromeda Nebula from a Spectroscopic Survey of Emission Regions." *The Astrophysical Journal* 159, no. 2 (February): 379–403. https://doi.org/10.1086/150317.

Rudwick, Martin J. S. 2014. *Earth's Deep History: How It Was Discovered and Why It Matters*. Chicago: University of Chicago Press.

Russell, Bertrand. 1928. *Sceptical Essays*. London: Routledge.

Russell, Bertrand. 1929. *Mysticism and Logic*. New York: Norton.

Sagan, Carl. 1997. *The Demon-Haunted World: Science as a Candle in the Dark*. New York: Ballantine Books.

Sandweiss, Martha A. 2009. *Passing Strange: A Gilded Age Tale of Love and Deception Across the Color Line*. London: Penguin Books.

Scharping, Nathaniel. 2016. "LHC Didn't Break Physics, New Particle Vanishes Upon Further Review." *Discover*, August 5. http://blogs.discovermagazine.com/d-brief/2016/08/05/physics-lhc-new-particle/.

Sciama. D. W. 1969. *The Physical Foundations of General Relativity*. Garden City, NY: Doubleday and Company.

Shanks, Niall. 2004. *God, the Devil, and Darwin*. New York: Oxford University Press.

Shapin, Steven. 1996. *The Scientific Revolution*. Chicago: University of Chicago Press.

Sheehan, William. 2003. "Secret Documents Rewrite the Discovery of Neptune." *Sky and Telescope*, July 23. http://www.skyandtelescope.com/astronomy-news/secret-documents-rewrite-the-discovery-of-neptune/.

Shubin, Neil. 2008. *Your Inner Fish*. New York: Pantheon Books.

Simpson, George Gaylord. 1944. *Tempo and Mode in Evolution*. New York: Columbia University Press.

Singham, Mano. 2000. *Quest for Truth: Scientific Progress and Religious Beliefs*. Bloomington, IN: Phi Delta Kappa Educational Foundation.

Singham, Mano. 2009. *God vs. Darwin: The War Between Evolution and Creationism in the Classroom*. Lanham, MD: Rowman & Littlefield.

Singham, Mano. 2011. "No Doubt." *New Humanist*, July–August, 32–33. https://newhumanist.org.uk/articles/2594/no-doubt.

Smith, Peter. 2013. *An Introduction to Gödel's Theorems*. Cambridge: Cambridge University Press.

Smolin, Lee. 2006. *The Trouble With Physics*. Boston: Houghton Mifflin.

Snow, C. P. 1964. *The Two Cultures: And a Second Look*. Cambridge: Cambridge University Press.

Soares, Domingos S. L. 2009. *Newtonian Gravitational Deflection of Light Revisited*. https://arxiv.org/pdf/physics/0508030v4.pdf.

Sollas, W. J. 1921. "Response to the Paper 'The Age of the Earth' by Lord Rayleigh." *Nature* 108, no. 2713 (October 27): 281–83. https://doi.org/10.1038/108281a0.

Soon, Chun Siong, Marcel Brass, Hans-Jochen Heinze, and John-Dylan Haynes. 2008. "Unconscious Determinants of Free Decisions in the Human Brain." *Nature Neuroscience* 11, no. 5 (May): 543–45. https://doi.org/10.1038/nn.2112.

Spencer, Herbert. 1852. "The Development Hypothesis." In *Essays Scientific, Political & Speculative*, Volume 1, 1–7. London: Williams and Norgate.

Stacey, Frank D. 2000. "Kelvin's Age of the Earth Paradox Revisited." *Journal of Geophysical Research* 105, no. B6 (June 10): 13155–58. https://doi.org/10.1029/2000JB900028.

Stachel, J. J., and Robert Schulmann (eds.). 1987. "Letter to Eduard Study from Albert Einstein dated Sep 25, 1918." In *Collected Papers of Albert Einstein*, Volume 8, 651. Princeton, NJ: Princeton University Press.

Stassen, Chris. 1998. "Isochron Dating." *The Talk Origins Archive*. http://www.talkorigins.org/faqs/isochron-dating.html.

Stuewer, Roger H. 1970. "Non-Einsteinian Interpretations of the Photoelectric Effect." In *Historical and Philosophical Perspectives of Science*, edited by Roger H. Stuewer, 246–63. New York: Gordon and Breech. http://mcps.umn.edu/assets/pdf/5.11_Stuewer.pdf.

Supran, Geoffrey, and Naomi Oreskes. 2017. "Assessing ExxonMobil's Climate Change Communications (1977–2014)." *Environmental Research Letters* 12, no. 8 (August 23): 084019. https://doi.org/10.1088/1748-9326/aa815f.

Swift, Jonathan. 1721. *A Letter To A Young Gentleman, Lately Enter'd Into Holy Orders*. London: J. Roberts.

Tan, Andi, Mengjiao Xiao, Xiangyi Cui, Xun Chen, Yunhua Chen, Deqing Fang, Changbo Fu, et al. 2016. "Dark Matter Results from First 98.7 Days of Data from the PandaX-II Experiment." *Physical Review Letters* 117 (September 16): 121303(1)–121303(7). https://doi.org/10.1103/PhysRevLett.117.121303.

Theocharis, T., and M. Psimopoulos. 1987. "Where Science Has Gone Wrong." *Nature* 329 (October 15): 595–98. https://doi.org/10.1038/329595a0.

Thucydides. 1910. *The History of the Peloponnesian War*. Translated by Richard Crawley. London: J. M. Dent, and New York: E. P. Dutton. http://classics.mit.edu/Thucydides/pelopwar.mb.txt.

Tolstoy, Leo. 1899. *What is Art?* Translated by Aylmer Maude. New York: Thomas Y. Crowell & Co.

Tremaine, Scott. 2011. "Is the Solar System Stable?" *Institute Summer Letter*, Institute for Advance Study. https://www.ias.edu/ideas/2011/tremaine-solar-system.

Weinberg, Steven. 2015. *To Explain the World: The Discovery of Modern Science*. New York: Harper Collins.

Wells, Jonathan. 2000. *Icons of Evolution*. Washington DC: Regnery.

Whitcomb, John C. Jr., and Henry M. Morris. 1961. *The Genesis Flood*. Philadelphia: Presbyterian and Reformed Publishing Company.

Index